Russia's Comparative Advantages in Foreign Trade

SCHRIFTEN ZUR
EMPIRISCHEN WIRTSCHAFTSFORSCHUNG
Herausgegeben von Peter M. Schulze und Peter Winker

Band 21

PL ACADEMIC RESEARCH

Ivan Savin

Russia's Comparative Advantages in Foreign Trade

Forecasting Potential Dynamics and Identifying Stimulation Measures

PL ACADEMIC RESEARCH

Bibliographic Information published by the Deutsche Nationalbibliothek
The Deutsche Nationalbibliothek lists this publication in the Deutsche Nationalbibliografie; detailed bibliographic data is available in the internet at http://dnb.d-nb.de.

Zugl.: Gießen, Univ., Diss., 2011

D 26
ISSN 1437-0697
ISBN 978-3-631-63993-1

www.peterlang.de

To my Mom and Dad, for their *unending love* and support.

Contents

Chapter 1

Introduction

The problem of defining industries with a higher potential for improving their performance on the domestic and foreign markets is always of interest for private companies and public authorities. To this end, the concept of comparative advantages is by all means one of the central instruments for measuring this performance. In the existing literature, comparative advantages are assessed either at a particular moment in the past or at present without any detailed forecast on their future dynamics. However, understanding factors that form those advantages allowing for countries to benefit from foreign trade is crucial both for the private and public sector willing to improve their economic strategies.

In particular, this information is highly relevant for economic policy making in economies in transition. For example, in Russia there is an ongoing political discussion on whether to support the automobile and aerospace industries. However, there is no clear justification with regard to their future comparative advantages. Furthermore, it is not clear what factors are most efficient to stimulate those industries. Since economic growth and innovations are widely seen to be interrelated (Schumpeter 1943, Porter 2003), the latter question transforms into search of instruments triggering innovative activity of companies.

The objectives stated require a well–founded theoretical analysis to be combined with a careful empirical investigation. In addition, a number of methodological improvements are necessary to mitigate certain limitations existing in the established research approaches. Motivated by the example of my home country, where most of comparative advantages are concentrated on hydrocarbons and other raw materials, I address those questions providing a suitable forecasting tool for comparative advantages' dynamics and investigating factors that stimulate innovations in the studies stated below.

The first Chapter **'Forecasting Russian foreign trade comparative advantages in the context of a potential WTO accession'** suggests a new approach called the Prospective Comparative Advantage index that forecasts comparative advantages' dynamics based on the relative price differences between countries. In particular, prospective comparative advantage takes place when the relative price on the domestic market is lower than on the foreign market. The method is efficient as long as (i) the trade relations of a particular country are not liberalized yet and (ii) the index is considered in conjunction with other available information (including, e.g., revealed advantages, exchange rate distortions) to control for public support.

To validate the approach, I apply it empirically to the Central and East European (CEE) countries, most of which had structural problems in their economies similar to Russia (Zaghini 2005), and show that industries in which these countries gained comparative advantages over the last two decades were undervalued. Furthermore, I identify prospective advantages for the Russian economy in some medium and high technological goods implying that in the course of time investments in these industries will be relatively more efficient.

In the paper titled **'Factor-biased technical change and specialization patterns'** I contribute to the literature on endogenous technical change (see, e.g., Romer (1990), Grossman and Helpman (1991), Aghion and Howitt (1992)) integrating technology–based with factor endowment–based views on trade in modelling effects resulting from capital–market integration. This study extends the Prospective Comparative Advantage index shedding light on the micro–foundations that explain the rapid growth of comparative advantages in technologically advanced industries in the CEE countries. Thus, in contrast to the existing literature, which argues that developing countries are anticipated to 'concentrate' on their comparative advantages (see, e.g., Acemoglu (2002)), I demonstrate why those advantages can be diversified.

In particular, integrating the approach of induced technical changes into a neoclassical Heckscher-Ohlin framework with a continuum of final goods, one can illustrate the mechanism why (i) comparative advantages can shift towards developing countries; (ii) companies in those countries invest in new technologies and thus (iii) contribute to a further expansion of related comparative advantages. Furthermore, accounting for state-dependant R&D processes, this study differentiates in time the effects of capital inflow on specialization patterns explaining the time lags in comparative advantages' responses observed empirically.

In addition, a series of measures for countries being in the period of transition (e.g., Russia) to realize their potential comparative advantages is discussed. In particular, effective measures on trade liberalization together with attraction of scarce production factors are meant. Since the market of technology innovations is potentially inefficient, this study also supports public measures to stimulate innovative activity of companies.

To analyze factors that trigger production of innovative goods and services in detail, those factors need to be estimated econometrically within a generalized model. However, as the number of hypotheses is large, the process of model selection becomes a crucial part of the empirical implementation. I demonstrate that the corresponding optimization problem is complex due to the large discrete search space. For this reason, heuristic optimization methods (see Gilli and Winker (2009) for an overview) are applied to Russian regional data within the framework of a log–linear dynamic panel data model in my second Chapter **'Heuristic optimization methods for dynamic panel data model selection: Application on the Russian innovative performance'**.

In particular, Threshold Accepting and Genetic Algorithms are employed. Based on Monte-Carlo simulation results, I show that both heuristic strategies can deal with the problem stated efficiently, whereas for a given computational time Genetic Algorithms provide marginally better results.

Estimating resulting model with the system GMM method (Blundell and Bond 1998), I find evidence supporting (i) adoption of foreign state–of–the–art–technologies, (ii) promotion of foreign direct investments, (iii) improvements in the education system and inter–regional knowledge spillovers. A series of possible measures to enhance the impact of those factors both from the side of public authorities and private sector is also discussed. Together with the two Chapters described above this study constitutes the core of my thesis.

Furthermore, this book is extended with two more studies. The first one titled **'Heuristic model selection for leading indicators in Russia and Germany'** presents an application of a heuristic algorithm to a vector autoregressive model forecasting industrial production in the two countries based on its past observations and corresponding leading indicators. Whereas there is a large body of empirical evidence for OECD countries that these models outperform univariate time series forecasts (e.g., Kholodilin and Siliverstovs (2006), Ozyldrim *et al.* (2010)), this is by far not the case for transition economies like Russia. Moreover, there is a lack of research examining the indicators' performance during the recent crisis. Therefore, this Chapter contributes to the existing literature in two ways: (i) by comparing the indicators with regard to their forecasting performance in Russia and Germany (ii) both prior and during the recent recession. The study benefits from the heuristic application that provides a more flexible form of the multivariate lag structure. I find that the leading indicators significantly improve produced forecasts during the crisis in Germany, but not in Russia. These results are confirmed by the conditional predictive ability test (Giacomini and White 2006) and are consistent with the Granger causality test and impulse response function analysis.

The last extension **'A comparative study of the Lasso-type and heuristic methods'** is devoted to a comparative analysis of heuristic subset selection methods with the so–called Lasso–type (Lasso, elastic net, adaptive Lasso) strategies (see, e.g., Tibshirani (1996)). While both methods have become increasingly popular for the model selection process in the last years, it was not clear which strategy provides superior results and what are necessary conditions for that. In this Chapter I address these issues. To this end, results of a series of Monte–Carlo experiments together with an application to an actual empirical problem (from Sala-i-Martin (1997)) are provided.

The remainder of the book presents a revised version of my PhD thesis and proceeds as follows. In Chapter 2 the Prospective Comparative Advantage index is suggested and applied to the CEE countries and Russia empirically. In Chapter 3 different factors triggering innovative activity of Russian companies are examined by means of the heuristic optimization methods. Chapter 4 presents an application of the heuristic optimization techniques to business cycle forecasting in Russia and Germany. In Chapter 5 a comparative study of the heuristic and Lasso–type model selection strategies is conducted. Chapter 6 concludes. The four studies constitute separate works and are presented as such.

The Chapters presented in this book can be furthermore found separately in the following journals:

Savin I. and P. Winker (2009). Forecasting Russian foreign trade comparative advantages in the context of a potential WTO accession, Central European Journal of Economic Modelling and Econometrics 1(2), pp. 111–138. http://cejeme.org/publishedarticles/2009-19-11-633908963415312500-1763.pdf

Savin I. and P. Winker (2012). Heuristic optimization methods for dynamic panel data model selection: Application on the Russian innovative performance, Computational Economics 39(4), pp. 337–363.
http://www.springerlink.com/content/e9923p30400740tn/

Savin I. and P. Winker (2012). Heuristic model selection for leading indicators in Russia and Germany, OECD Journal: Journal of Business Cycle Measurement and Analysis 2012/2 (December 2012). http://www.oecd.org/std/jbcma

Savin I. (forthcoming). A comparative study of the Lasso–type and heuristic model selection methods, Journal of Economics and Statistics (forthcoming). http://www.jbnst.de/en/content.php

In comparison to the version of the thesis submitted to the Justus Liebig University of Giessen on May 2, 2011 paper titled 'Factor–biased technical change and specialization patterns' by Jana Brandt, Jürgen Meckl and myself is not included in this book, but can be found in MAGKS Joint Discussion Paper Series in Economics 18-2011 at http://www.uni-marburg.de/fb02/makro/

forschung/magkspapers/18-2011_savin.pdf. In addition, Chapters 5 and 6 have been changed based on the comments and suggestions from anonymous referees from the respective journals.

Chapter 2

Forecasting Russian Foreign Trade Comparative Advantages

in the Context of a Potential WTO Accession

Ivan Savin, Peter Winker
The final publication is available at http://cejeme.org/

2.1 Introduction

The problem of defining industries with a higher potential for improving their performance on the domestic and foreign markets was always of interest for private companies and public authorities. In the last decades the concept of comparative advantages (CAs) became the standard instrument for measuring the performance. Apparently, the existing instruments of the comparative advantages theory assess dimensions of advantages at a particular moment in the past or at present, whereas a CA is a dynamic characteristic, i.e., it changes in the course of time according to the economic development of countries (Grossman and Helpman 1991).

For Russia this issue is even more important due to the transition period in its economy and the necessity to set development priorities. The modernization of the economy using newest technologies is the core of Russia's long term economic strategy. To achieve this goal the government organized several programmes (e.g., 'Federal Targeted Programme of Nanotechnologies') to stimulate innovative activities of enterprizes and diversify thr export structure by gaining CAs in industries with high value added goods. Therefore, an instrument that could tackle the problem of forecasting CAs is required to identify industries, which have a higher potential in improving their performance and, therefore, are more attractive for private and public investments.

At the moment this area of research is seldom explored and contains only a few scientific publications. For example, Belousov (2006) based on macroeconomic trends suggests that the future CAs of Russia could be energy, transport and the agriculture sector. Rogov (2004) argues that a diversification of Russian foreign trade is possible in high technology industries based on public corporations as the most effective management instrument without providing any concrete suggestions for industries where CAs will take place. Therefore, so far as we know research in this area relies on intuitive, ad hoc approaches rather than theoretical analysis and quantitative assessments, and our research is intended to close this gap.

This paper is organized as follows. Section 2.2 provides an analysis of the comparative advantages theory and different approaches on measuring CAs. We present a new method of forecasting CAs' dynamics in the form of a 'prospective' comparative advantage index and describe main features of this index. In Section 2.3 the present state of the Russian foreign trade specialization is explored and the index of prospective comparative advantages is empirically tested on the example of new EU member states. Based on these results we identify industries with apparently higher potential for formation of CAs in Russia. Section 2.4 concludes.

2.2 The Concepts of Comparative Advantages

2.2.1 Revealed Comparative Advantages

In order to measure the current trade specialization of a country and its potential in the context of trade liberalization, we consider CAs' indices. According to the principle of comparative advantages, trade patterns and export specialization are determined by CAs that are difficult to assess empirically, since the principle is based upon autarkic prices which we can not assess under free trade (Balance *et al.* 1987). There exist numerous alternatives to measure comparative advantages (Bruno (1965), Krugman and Hatsopoulos (1987), Porter (1990)). Balassa (1965) suggested calculating the Revealed Comparative Advantage (RCA) index as follows:

$$\mathrm{RCA}^B_{ij} = (x_{ij}/X_{wj})/(x_i/X_w), \tag{2.1}$$

where

x_{ij} export of good j by country i
X_{wj} world export of good j
x_i export of all goods by country i
X_w world export of all goods.

If the Balassa index RCA^B_{ij} is greater (smaller) than 1 country i exhibits a revealed comparative advantage (disadvantage) in good j. With other words, Balassa argued that country i exhibits a comparative advantage in good j, if its export of good j, relative to world export of good j, is larger than the country's market share in total exports.

Balance *et al.* (1987) classified different approaches to measure revealed CAs and tested their consistency. The authors highlighted two main classes of indices, namely those using data on trade, domestic production and consumption; and those using only trade data. The consistency tests based on the comparison of correlation coefficients for alternative pairs of RCA indices, rankings of countries provided by the RCA indices and the extents to which the indices agree in distinguishing between countries that enjoy CAs, showed that the index based on the 'net export' approach (that uses only trade data) is the most reliable among all tested indices. Taking this into account, a more appropriate index of CAs would be as follows:

$$\mathrm{RCA}^N_{ij} = 100(x_{ij} - m_{ij})/(x_{ij} + m_{ij}), \tag{2.2}$$

where x_{ij} and m_{ij} denote the export and import of good j by country i, respectively. The denominator in (2.2) represents total trade volume of good j of country i (Balance *et al.* 1987). This index is ranging between -100 and 100. In practice the absolute value of the index is rarely larger than 10. The larger the index is, the higher is the foreign trade specialization in industry j.

Another approach for CA assessment was proposed by Lafay (1992) and takes intra–industry trade more closely into consideration. This is particularly important in the context of the globalization of the world economy, the growth of semi–finished goods trade flows and multinational companies (MNC) that distribute the production process among their subsidiaries around the world. The Lafay index (LFI) also has an advantage in comparison to (2.2) as it is robust to macroeconomic distortions, since it takes the difference between each item normalized trade balance and the overall normalized trade balance into account (Zaghini 2005).

The Lafay index can be calculated for a given country i and for a given good j as follows:

$$\mathrm{LFI}_{ij} = 100\left(\frac{x_{ij}-m_{ij}}{x_{ij}+m_{ij}} - \frac{\sum_{j=1}^{N}(x_{ij}-m_{ij})}{\sum_{j=1}^{N}(x_{ij}+m_{ij})}\right)\frac{x_{ij}+m_{ij}}{\sum_{j=1}^{N}(x_{ij}+m_{ij})}, \qquad (2.3)$$

where N is the overall number of traded goods. So the Lafay index measures the contribution of each good to the overall normalized trade balance. Therefore, the sum of LFI indices for all goods equals zero. Positive values of the Lafay index indicate the existence of comparative advantages and vice versa.

In general we can conclude that the indices of revealed comparative advantages assess advantages that exist and are realized in practice, but give no information in regard to their future dynamics.

2.2.2 Prospective Comparative Advantages

It is both for the private and public sector in any particular economy crucial to know, what industries may exhibit CAs, which for some reasons are not realized. This can efficiently help all current and potential actors to improve their economic strategy. Since CA tends to change over time due to changes in technologies or factor endowments, a forecasting instrument of CAs dynamics is required. Looking for an appropriate instrument we realized that the existing approaches are mainly based on the investigation of reasons for cost advantages and normally do not provide any detailed estimation of industries with a potential of CA.

The widely cited work of Grossman and Helpman (1991) noted that dynamic comparative advantages arise 'due to incipient cost advantage in R&D'. The authors devote their attention mostly to knowledge spillovers and its accumulation, as well as to the relative shares of R&D expenditures in GDP, which are, in fact, instruments for stimulating comparative advantages. Our intent is to identify the industries, where these instruments may be most advantageous.

As for empirical based analysis, the existing research (Marconi and Rolli (2008), Landesmann and Stehrer (2001), Montobbio (2003)) provides estimates of the potential impact of explanatory variables based on CA dynamics in the past, but do not enable to forecast CA dynamics in the future, mainly due to high variance across country specific characteristics and great difference of CA determinants across industries depending on their level of technology.

Here it is reasonable to note that we will use the classification of high, medium and low technology industries defined by OECD (1986). Therefore, industries with a share of R&D in output larger than 4% are characterized as high technological industries (e.g., pharmaceuticals, microelectronics, aircraft), industries with a share of 1-4% are considered as medium technological industries (chemicals, motor vehicles, non-electronical machinery) and all other are categorized as low technological industries (mining industries, textile, food).

Marconi and Rolli (2008) examined 16 emerging countries over the period of 1985-2000 and showed that higher unit labor costs have a negative effect on CAs, whereas intensive physical capital accumulation and imports of capital goods have a positive effect only on medium and high technological industries. Similarly, Montobbio (2003) showed that R&D per employee positively affects export market shares, whereas industries are affected by R&D with different time lags, e.g., high technological industries are affected by R&D with a time lag of three years and medium technological industries are affected with a time lag of one or two years.

Stehrer and Woerz (2003) have determined and distinguished three types of 'catching-up' processes in productivity levels of less developed countries: first, the 'continuous–convergence approach' with an equal speed of convergence across all industries; second, the 'climbing–up–the–ladder approach' where the lagging country closes the gap in the low technological industries before it can start to catch up with high technological industries and, third, the 'jumping-up approach' where the lagging country catches up in the high technological industries from the very beginning. The authors have empirically tested these 'catching-up' patterns and concluded that these patterns differ substantially across countries in terms of industries, time and rate of growth of the 'catching-up' process.

To forecast the dynamics of CAs, we will consider relative prices of products in different countries. Practically, we turn back to the initial idea of the comparative advantages theory: when two countries have different prices of goods in autarky, and trade is allowed to be opened up, these countries gain CAs in goods, which are less expansive than in another country. Thus, a CA is established by cost comparison under autarky and under trade (Siggel 2006).

Possible reasons for this cost advantage include: differences in countries′ endowments of factors of production, technologies, taxation, and consumers' preferences. The most popular theory of comparative advantages is the Heckscher-Ohlin theory. This theory describes differences in factor proportions among countries as a main reason of CAs. For further details on the comparative advantage theory see Dixit and Norman (1989).

The principle of comparative advantage is straightforward as long as only two products are involved. The extension to a continuum of goods demonstrated by Dornbusch *et al.* (1977) confirms the principle as valid and determines a 'borderline', according to which a country has a CA in goods, that it can produce with less production costs than other country. Further, Dornbusch *et al.* (1977) extend the analysis by requiring the usage of equilibrium prices to measure cost differences across countries as non-equilibrium prices contain various distortions like trade barriers and exchange rate misalignments. Thus, cost comparison based on market prices can not be the sole basis of a measure of CAs. When costs are measured in terms of distorted prices we deal with competitive advantage, which is not the same as CA; having a CA implies that production costs in terms of equilibrium factor prices are lower than those of an international competitor.

The difficulty in this case is that, as mentioned above, we cannot measure autarky prices simply because we can not observe them (Balance *et al.* 1987). An important assumption that we make here is that we will consider commodity prices in countries under numerous trade barriers as approximations to autarky prices that will converge to international prices after economic liberalization. Therefore, this forecasting instrument is suitable only for countries that exhibit rigid protectionist policies. In addition, these countries are assumed to be in the period of transition and are expected to liberalize their economy in terms of joining multinational free trade zones, e.g. WTO, EU. Here we should take into account that the Russian Federation is the largest economy outside the WTO agreements and from this point of view may rather be considered, at present, as a closed economy.

Let us now turn back to the meaning of the differences in commodity prices across countries. When a country has a CA in a particular good, it can produce an additional unit of this good while refusing to produce less units of other goods than an other country or the world economy in general. Therefore, the relative price of this good is also lower in this country than elsewhere (Dixit and Norman 1989). In the case of an open economy, the country will export this good according to the Ricardo principle. This good then becomes scarcer on the domestic market and, therefore, more expensive until it will be as expensive as on the foreign one. This is exactly the comparative advantage principle that is described in classical and neo-classical trade theory (Siggel 2006).

But when trade barriers exist, this process is limited and domestic prices do not reach the level of international prices and some natural CAs are not realized because of distorted prices. Therefore, based on relative price differences we can conclude whether there is a prospective, but unrealized CA or not.

As we have just argued, it is difficult to measure CAs based on market prices due to various distortions presented by these prices (Siggel 2006). To minimize these distortions we use relative producer price indices. This has several reasons, which we will now discuss in detail.

First, since we compare advanced countries and countries in transition with a different structure of industrial output as compared to OECD countries, we use price indices rather than absolute prices. This will eliminate the need to translate the prices in one currency and will better suit our model of heterogenous goods and levels of economic development. Moreover, compiled at different levels of aggregation, the indices are intended to be free of some of the defects of existing measures and more suitable for the calculation of changes in quantities traded and for the analysis of the relations between prices and quantities (for further details see Diewert (1993)).

Second, the usage of relative prices to construct an indicator of an industry's competitiveness was realized long ago. Indicators of competitiveness calculated by the OECD's Economics and Statistics Department are based on the industry consumer price indices (Durand and Giorno 1987). Practically, they use a relative industry price (relative to foreign competitors), with the exchange rate translating it into US dollars.

Third, in contrast to Durand and Giorno (1987), we consider manufacturer prices rather than consumer prices to measure industry competitiveness, because they are less distorted by the state of competition on a particular market and by possible misalignment of the exchange rate. In fact, producer prices are actual transaction prices of domestic production taking account of discounts and rebates, excluding VAT and all taxes and duties on the goods as well as trade margin. As it was shown by Nakamura and Steinsson (2008) trade margin is stochastic and not fixed as a percentage from a producer price and accounts for roughly 50% of consumer price instability. So we consider that manufacturer prices reflect changes in technologies in companies to a larger extent than consumer prices that are not well suited to measure competitiveness indices (Siggel 2006).

Durand and Giorno (1987) argue that the producer price indices vary in quality across countries due to their lack of homogeneity in terms of weighting and coverage. They argue that the main problem of using manufacturer prices is the difference in data collection across countries, in general, and a possible omission of some products from countries′ price indices, in particular. It is true, that the problem of economic data aggregation, in particular data aggregation on prices (Balk 1983), is very complicated. But we must keep in mind that, in first instance, we suggest using producer prices with a fine level of commodity detail (classification of goods of 3- and 4-digits) rather than on the high level of aggregation, at which the criticism is primarily directed and, in second instance, we propose to use (if possible) a unified database, e.g. Eurostat, for all countries in the scope as a source of industrial output prices to minimize the risk of omitted goods.

Thus, we use an indicator of competitiveness, or in other words 'competitive advantage', as a forecasting instrument for CAs in the context of economic liberalization. Here we need to admit that it is very unlikely to eliminate all artificial distortions that take place in the real economy, e.g., exchange rate misalignments. But it is beyond the scope of this paper to approximate 'true natural' CAs. We rather try to forecast CAs that take place in the real economy with a minimum of trade barriers and can be measured by indices of revealed CAs like the Balassa index based on the trade statistics (see Siggel (2006)).

We call the index that forecasts CAs based on relative price differences the Prospective Comparative Advantage (PCA) index. Thus, prospective advantage takes place when a country has a CA, but it is not exploited (e.g. because of trade barriers). The PCA index is defined as follows:

$$\mathrm{PCA}_{ij} = \frac{p_{it}^{h}}{p_{jt}^{h}} \Bigg/ \frac{p_{it}^{f}}{p_{jt}^{f}}, \tag{2.4}$$

where

p_{it}^{h} price index of good i on the domestic market in period t
p_{jt}^{h} price index of good j on the domestic market in period t
p_{it}^{f} price index of good i on the foreign market in period t
p_{jt}^{f} price index of good j on the foreign market in period t

We denote a base good as j. In our case it is an overall producer price index (including good i). Prospective comparative advantage takes place when the relative price on the domestic market is lower than on the foreign market. An index value less than 1 reflects a prospective comparative advantage in this industry. The lower the index is, the higher is the prospective comparative advantage.

The PCA index may be seen as an extension of the Domestic Resource Cost (DRC) criterion proposed by Bruno (1965). This criterion is known as an alternative measure of revealed comparative advantages. The main difference is that instead of measuring the relation of total domestic costs per unit of a product to free-trade price of comparable output in absolute values, we measure producer price indices of goods in relation to the base good among two countries. Measuring competitiveness among countries under trade barriers and presuming economic liberalization in the nearest future, the PCA index provides a forecast of industries in which CAs are likely to arise rather than assessing CAs that are already realized.

Calculating the PCA for Russia we can consider Germany, the biggest trade partner of Russia, or the European Union as a whole as the foreign economy, which can be considered as liberalized economies and with price indices structures close to the world economy. The choice of the EU as the foreign economy is also supported by the fact that the EU has the highest degree of product coverage in its trade pattern with Russia (Brenton *et al.* 1997). It is also important to note that the more similar the industrial structure of the foreign economy is, the more exact is the PCA index due to the comparability of overall producer price indices. For Russia and the EU it is not exactly the case due to a significant prevalence of mining industries in the Russian economy as we will see in Section 2.3 of this paper.

2.2.3 Implications of the Prospective Comparative Advantages Index

If the PCA is lower than 1, we may conclude that industry i is undervalued in comparison to the overall industrial output, i.e. prices in a particular industry grow slower than the overall price index in comparison to a foreign country, and when the PCA is higher than 1, industry i is overvalued in comparison with the overall output, i.e. prices in this industry rise faster than the overall price index as compared to the foreign economy. After economic liberalization an inefficient use of resources will hardly be maintained, prices in undervalued industries will most likely rise and the optimal strategy would be to arrange conditions under which resources from overvalued industries could freely float into undervalued industries.

To make it easier to understand, let us take a concrete calculation example: consider Russia for h and Germany for f. We will consider medical and surgical equipment for i, an overall industrial price index for j and will calculate the PCA index in 2007 (t). We use chain indices, so that price indices of 2006 are all assumed to be equal 100. Thus, p^h_{it}= 102,1 (which means that the producer price for medical and surgical equipment has changed slightly in Russia in 2007 in comparison with 2006), p^h_{jt}= 125,1 (the manufacturer prices in Russia have risen dramatically in 2007), p^f_{it}=100,1 and p^f_{jt}=100,4 (the prices in Germany of an overall industrial output as well as of medical and surgical equipment in particular have not risen much, at least in comparison with Russia). Then, according to (2.4) the PCA equals 0,823 and gives a signal that this Russian subindustry is undervalued, prices in this industry are likely to rise after economic liberalization and, therefore, this industry is prospective for investments.

Measuring the PCA indices has several obstacles. First, the PCA is a dynamic characteristic. Therefore it may be distorted, e.g. by price bounces. Second, the PCA is dependent on initial conditions. For example, if the price of a good initially equaled 1 and has risen to 3, then the growth rate is 300%, but if the price initially equaled 2, then the growth rate is only 50%. So as we see, a change of 50% in initial conditions has led to a more significant change in the PCA. It may therefore become problematic to make conclusions based on a quantitative comparison of PCA indices. Third, this index is effective when the share of traded goods (international trade) is small relative to GNP, i.e. the closer is an economy to autarky, the better we can measure autarky prices and the more informative is the PCA index. Finally, the PCA is also influenced by the exchange rate. When a currency is undervalued, it stimulates net exports and reduces the share of international trade in GNP. Imported goods become more expensive (in terms of p_{it}^{h} in (2.4)) on the domestic market than on the foreign market and this will artificially reduce prospective comparative advantages, e.g., industries may be estimated as overvalued, when in fact they are not. Therefore, in cases where exchange rate distortions are present, the actual PCA value need to be assumed higher than the value of the index. It is not to oversee that an undervalued exchange rate also affects revealed comparative advantages (RCA) but in the opposite direction: it stimulates export of goods, which are not characterized as advantageous for a particular country.

Thus, we need to admit that the measurement of the PCA index as of any other index of competitive advantage is in a large measure a matter of trade-off with available data. In addition, some technical considerations, e.g., collecting and matching the price indices with different classifications of goods, arise in the calculation of the prospective advantages index.

The PCA index should not be considered isolated from other economic indicators, in particular from the indices of revealed comparative advantages, e.g., the Lafay index. This is because of various factors that may lead to over- and undervaluation of prices in a particular industry. After all, we consider not autarky economies, but economies with high trade barriers that are still able to trade. Thus, undervaluation can be explained not only by lower production costs in a given country, but also by intensive competition on the market. Overvaluation in turn can result from favorable international prices. For a summary see Table 2.1.

Therefore, to explore whether a prospective CA is presented in a particular industry or not we need to consider the PCA index in conjunction with other instruments like RCA index.

Table 2.1: Revealed and Prospective comparative advantages

	Revealed Comparative Advantage	Revealed Comparative Disadvantage
Prospective Comparative Advantage	A country has an advantage in this good, but its export potential is not used in full. Trade barriers may discriminate imported semifinal inputs as well as prohibit to realize the full export potential of the country. Therefore, a growth of relative prices in this industry is expected, and this industry is attractive for further investments.	The undervaluation is temporary due to e.g. intensive price competition on the market between domestic and foreign players. The country may have a CA in this good, if this is an infant industry, and foreign goods are still dominant on the market that explains the revealed disadvantage. Otherwise, foreign producers have CA, but because of government support, domestic production is presented on the market. But in the course of economic liberalization resources from this industry will eventually flow to other industries.
PCA =1	The export potential is fulfilled.	There is a stable comparative disadvantage in this good.
Prospective Comparative Disadvantage	The overvaluation can be assumed as temporary because of high international prices in this good (e.g. petroleum). When international prices will go down (stabilize), the overvaluation will decrease as well. In this case a country may have an advantage in this good or not. If the effects of undervalued exchange rate and import barriers prevail, export is artificially induced and there will be no advantage in this good. But if there is a real cost advantage in producing a particular good, then the country has an advantage in this industry.	The country has no advantage in this good and the import share in this good is likely to rise in the future.

2.3 Empirical Application

2.3.1 Data Description

To compute the Revealed Comparative Advantage index for new EU member states and the Russian Federation we use data from the COMTRADE Database of the United Nations Statistics Division (UNSD) at the Harmonized System (HS), 2-digit, and for some particular industries at the 4-digit level covering the period of 2002-2006. These data are available in US dollars.

The data on the LFI indices for new EU countries in the period of 1993-2001 that we use in our empirical estimation were kindly provided by Andrea Zaghini (Center for Financial Studies, Frankfurt).

One of the problems in calculating the Prospective Comparative Advantage index of a country group including CEE countries and Russia is, first, to obtain data for the producer price indices with a detailed classification of industries and sub-industries for a period from the beginning of the 1990s.

Unfortunately, it turned out to be impossible to find detailed price indices of these countries for the period 1993-1996. Therefore, we will calculate and compare PCA indices for the period 1997-2004. Moreover, for the Czech Republic, Hungary and Slovenia we can calculate the PCA index only since 2001-2002 due to data limitations. Nevertheless, we need to keep in mind that the EU enlargement took place only in 2004 and the Central European Free Trade Area (CEFTA) and the Baltic Free Trade Area (BFTA) formed by these countries at the beginning of the 1990s have abolished duties on most of the industrial goods with the European Union only in 1997 (Adam *et al.* 2003). Therefore, we presume that by 1997 the relative prices were not fully adjusted yet and calculation of the PCA index for the period 1997-2004 will provide us with meaningful estimates.

Due to data limitations for producer price indices for new EU countries we use the harmonized indices of consumer prices (HICP) to calculate the PCA indices. As a consequence we are able to compare relative prices mostly on consumer products and not on the semi-finished goods. The data covers the period of 1997-2007, but with substantial differences in availability across countries and industries.

For the EU countries we use data from the Statistical Office of the European Communities (Eurostat) Database. For the Russian Federation we use data for the consumer and producer price indices from the Russian Federal State Statistics Service (Rosstat). Calculating the PCA index for Russia based on consumer prices we are forced to match two different classifications of goods and services, one is COICOP provided by Eurostat and another one of the Russian Rosstat. Calculating the PCA index based on the producer prices we have the same problem by matching the NACE classification for the EU countries and the classification of Rosstat. All three datasets have a detailed classification of goods, at the 2-, 3- and 4-digits. We consider that these classifications of goods are similar and conclude that the matching will not seriously affect our results.

2.3.2 The Example of the New EU Member States

To test the mechanism of Prospective Comparative Advantages we use an example of the new EU member states, in particular Poland, the Czech Republic, Hungary, Slovenia, Slovak Republic, Cyprus, Malta, Lithuania, Latvia and Estonia. These countries may be easily split in two groups. The first group is formed by Poland, the Czech Republic, Hungary, Slovenia, Slovak Republic and the Baltic states that were members of the former socialist bloc. So the structural organization of their economies at the beginning of the 1990s may be considered as close to the system that Russia had at that period of time. The remaining countries, Malta and Cyprus, were taken to complete our analysis with a different type of economies that have never been planned economies. We have chosen these countries because they have recently passed transition periods of their economies. In addition, most of these CEE countries as well as Russia have a larger share of skilled-labor force in comparison to other developing countries (Bardhan and Kroll (2006), Zaghini (2005)). So we can draw a parallel between initial conditions that these countries had at the beginning of the 1990s and conditions in which Russia is situated at the moment.

According to Zaghini (2005) most of these countries had excessive natural, labor and land resources in comparison to other EU countries, so it is not surprising that in the period of 1993-1994 they mainly specialized in sectors where these resources are used more intensively, namely products of steel and glass, simply worked wood, vegetables and fruits. Zaghini (2005) showed that these countries managed to gain CAs during a period of ten years in which they were weak at the beginning of the liberalization process, notably in some medium and high technological industries, e.g., transport and machinery building, electronics. The author explains this success by the 'advantage of backwardness' (Gerschenkron 1962) together with significant investments both from the side of governments and from EU countries and by substantial technological transfer as a part of the FDI inflows that was successfully adopted thanks to the endowment of a skilled labor force.

The realization of this scenario could be a reasonable policy in purposing the Russian Government objectives until 2020. Understanding what was the general characteristic of these industries in CEE countries will help us to reveal prospective industries for investments and stimulating CAs in Russia. Our idea is that these 'successful' industries were undervalued (had prospective advantages) and thanks to the inflow of new technologies and investments became realized.

Zaghini (2005) used the Lafay index (LFI) and the world export share (WES) to assess CAs at the beginning of the liberalization period (1993-1994) and after

7 years of reforms (2000-2001). For example, the Czech Republic specialized in simple products of glass, iron and wood with a minor specialization in passenger motor cars in the period of 1993-1994. But by 2001 it has trebled its world export share in motor vehicles and significantly enlarged its revealed advantages not only in this industry, but also in related industries like parts and accessories of motor vehicles, and also in electrical machinery.

Our idea is to calculate the PCA index described in the first part of this paper for industries, in which these countries have gained substantial CAs and to evaluate whether our index has 'forecasted' future success of these industries. We calculate the PCA index based on consumer price indices for new ten EU member countries using the EU as a foreign economy. Thus, we expect to find that the 'successful' industries were undervalued in terms of relative prices in 1997 and these prices have converged to the international free-trade level (the level of the EU) by 2004.

As we found out, based on the PCA index Estonia, Lithuania, Latvia, Slovak Republic, Poland and Malta had prospective advantages in telecommunication, furniture, clothing, electrical equipment, transport and petroleum industries in 1997. These industries got revealed CAs in the trade structure of these countries by 2004 according to the LFI index.

We also need to note that these undervaluations which reveal prospective advantages are minor in absolute values and normally range between 0.9 and 1 in terms of the PCA index. These minor undervaluations can be explained by measuring the indices for 1997 instead of 1993. During the period of 1993-1997 the economic liberalization of these countries was in progress and we may assume that the process of price convergence has already started as well. By 2004 the relative prices generally achieved the EU level, which corresponds to the value of 1 of the PCA index. The obtained results on the PCA index for the CEE countries are presented in Figure 2.1 and Table 2.2.

Table 2.2: An example of the PCA index for Poland.

	COICOP classification	1997	2004
cp0511	Furniture and furnishings	0,92	0,99
cp03	Clothing and footwear	0,94	1,01
cp031	Clothing	0,95	1,01
cp0311	Clothing materials	0,93	1,00
cp0911	Equipment for the reception, recording and reproduction of sound and pictures	0,95	1,02
cp0914	Recording media	0,93	0,99

Due to limited data availability we can calculate the PCA indices for Slovenia, the Czech Republic and Hungary only since 2001-2002. Thus, for example,

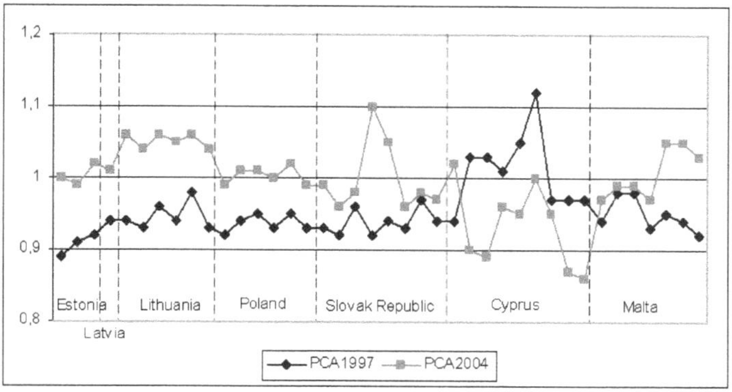

Figure 2.1: PCA indices for 1997 and 2004 for seven CEE countries.

in Hungary there were no prospective advantages to be observed in telecommunication or automatic data processing industries by 2002. We can assume that these industries are flexible in relation to price changes and adjustments. In contrast, production of meat and the motor car sector were undervalued even in 2002. We have calculated that in these two industries the indices of LFI and the world export share have generally increased by 2006 up to 1.01 and 0.7 respectively in passenger motor cars - and up to 0.28 and 0.95 in production of meat. It means that the prospective advantages that Hungary still had by 2002 were realized by 2006. Thus, we can draw a parallel with the 'catching-up' process observed by Stehrer and Woerz (2003): the speed of convergence in price levels between developed and developing countries that leads to changes in trade specialization patterns differs across industries, where this convergence process takes place.

A different example is shown by Cyprus. In this country the relative prices have significantly dropped after joining the EU, below the European relative price level. As an example, the relative prices of fruits and vegetables have decreased in terms of the PCA index from 1.05 and 1.12 to 0.95 and 1.00, respectively. This overvaluation can also be observed on Figure 2.1. As a result, the LFI and the world export share for fruits as well as for vegetables have decreased. This may indicate that overvaluation by the end of the 1990s was caused by high import duties in Cyprus before joining the EU. In fact, in comparison to other European countries before their accession to the EU Cyprus had high import tariffs (average tariff 37.6 percent) especially on agricultural products (with average rates above 60%) together with substantial export subsidies. It also should be noted that Cyprus is a fairly small economy, a net importer of manufacture goods, with 70 percent of GDP and foreign exchange receipts generated in services (WTO (1997)).

Having only very fragmented data on producer price indices we have calculated the PCA index that supports the results we have obtained on the consumer prices. Thus, the Czech Republic had prospective advantages in transport equipment and Lithuania had prospective advantages in production of outer garments and electrical machinery and equipment in terms of manufacturer prices as well. Since the data on producer price indices is very fragmented for CEE countries, we can not provide more detailed results.

The fact that we have got consistent results of PCA index for the CEE countries based both on producer and consumer price indices can be explained by changes in exchange rate regimes among these countries in 1997. In this period the exchange rate regimes of these countries, which differed considerably at the beginning of the 1990s, became oriented towards the Euro (Backe 1999). Therefore, possible distortions of the PCA index due to exchange rate misalignments have not significantly affected the results.

To assess the forecasting power of the PCA index we tested it using the following equation:

$$\mathrm{LFI}_{ij}^{diff} = \alpha + \beta \mathrm{PCA}_{ij}^{1997} + \varepsilon, \tag{2.5}$$

In this equation, LFI^{diff} stands for the absolute change (increase or decrease) of the Lafay index in the period from 1997 to 2001. This was done to measure the assumed effect of the PCA and to minimize the country character specialization in particular industries. Thus, Slovak Republic in 2001 has a Lafay index in passenger motor cars of more than 6, but the real improvement of the index during the period of 1997-2001 was about 4. PCA^{1997} is the calculated PCA index for 1997. We test (2.5) on industries with a 3-digit level on 7 countries: Poland, Slovak Republic, Latvia, Lithuania, Estonia, Cyprus and Malta (for which price indices are available since 1997). These industries were chosen independently from the fact whether a change in the LFI index is to observe in the period or not, according to only one criterion, namely data availability.

The results are presented in Table 2.3. Firstly, we have estimated (2.5) on goods from low, medium and high technological industries (total sample). Then, we have estimated (2.5) only on medium and high technological goods (subsample 1). Finally, we have also tested (2.5) only on industries, where a change of the LFI index was larger than 0.5 in absolute value (subsample 2). The subsample 2 was constructed to test, whether a significant difference in estimates compared to the total sample exists or not.

Table 2.3: Regression results.

	Total sample	Subsample 1	Subsample 2
β	-9.98	-15.79	-11.04
Std. Error	0.74	1.78	1.29
P-value	0.000	0.000	0.000
R^2	0.48	0.49	0.49
N	200	81	74

We have also tested a modification of (2.5) using logarithm of the PCA index and obtained almost the same results as for absolute values of the PCA index. This is due to the type of the PCA values, i.e. around 1, normally in the range from 0.8 to 1.2.

In addition, we have tested (2.5) controlling for possible country-specific effects using dummy-variables. As a result, no significant country-specific effects have been obtained.

Looking at Table 2.3 we see that the explanatory power of the index remains almost the same for all three samples (which is approximately 49% and quite well for a cross-section analysis with only one explanatory variable). The value of β - which is in all cases significantly different from zero - differs across the subsamples. β is substantially higher for the industries with a significant change in the LFI value, and even higher for the subsample of medium and high technological products. It means that in the case of medium and high technological products a negative/positive change in the PCA index of 0.1 (10% higher under- or overvaluation of prices) leads to an increase/decrease in the value of the Lafay index of roughly 1.6, i.e. industries that were undervalued in 1997 experienced a growth in CAs. This negative correlation of the change in the LFI value and the value of the PCA index can be observed on Figure 2.2 in Appendix 2.5, where LFI and PCI indices are grouped by country.

To conclude, we observe a robust significant influence of the PCA index on the CAs dynamics.

2.3.3 Russian Comparative Advantages

Revealed Comparative Advantages

Let us now turn back to the comparative advantages theory and consider dynamics of CAs indices of Russia during the last years. The RCA indices analyze the current structure of international trade. Table 2.4 in Appendix 2.5 presents a short review of obtained revealed comparative advantages of Russia in the form of LFI indices, starting from industries with the highest advantages and presenting the most disadvantageous industries in the end of the table.

Thus, we see that Russia's main CAs are linked to hydrocarbons (petroleum, gas, coal) and some other resources (woodworking industry, non-ferrous metals and fertilizers). Medium and high technology industries are characterized mainly as disadvantageous in Russian exports.

Moreover, in contrast to most of the developing countries whose trade structures have become more diversified in the last decade (Woerz (2005), Marconi and Rolli (2008)) the structure of Russian foreign trade has become even more concentrated in the period from 2002 to 2006. For example, in pharmaceutical products and motor cars the revealed disadvantages have significantly increased. The only exceptions among medium and high technology industries are nuclear reactors and turbo-jets, which are characterized as advantages, but very specific ones.

Prospective Comparative Advantages

As it was already pointed out in Section 2.2, the PCA index based on manufacturer prices is less distorted. Nevertheless, analyzing the prospective CAs of Russia we consider the PCA index based both on consumer and producer prices to compare the results and to try to explain possible deviations.

Calculating PCAs for Russia, we need to remember that according to the existing Exchange Rate Deviation Index (ERDI) the Russian rouble is still clearly undervalued (Tiusanen 2007). This index of undervaluation was extreme after the economic crisis in Russia in 1998 (roughly by factor 3) and has gradually decreased during the last ten years to about 1.8. This leads to a competitive advantage of Russian companies in terms of absolute prices. That points out that Russian companies have a competitive edge and an import potential in Russia may not be full exploited, i.e. the domestic market is protected by the undervalued rouble. Therefore, the undervaluation of prices measured by the PCA index may not be as significant as it really is, the PCA index based on the consumer prices exhibits a significant exchange rate distortion.

In Table 2.5 in Appendix 2.5 we see that based on the consumer prices prospective advantages of Russia contain motor cars, tobacco, medical and pharmaceutical products. Some moderate prospective advantages can be also found in clothing materials, whereas clothing industry itself is characterized with no advantages.

To deepen our prospective advantages analysis we will consider producer price indices as a base for the PCA mechanism, as these indices present production costs at a larger extent. Owing to a lack of data for the European Union as a whole, calculating the PCA we will consider Germany, the biggest trading partner of Russia, as the foreign partner (see Table 2.6 in Appendix 2.5).

As we see the obtained results differ from the PCA indices based on the consumer price index, i.e., producer prices′ volatility between the countries is more significant than volatility in consumer prices. One of the reasons for this may be that consumer prices between different markets are closely connected. Thus, an increase in prices in Germany on motor vehicles will most likely cause an increase in prices in Russia as well, and not only on imported cars, but on most cars, because other car sellers will react to the changes on the market. In contrast, a similar increase in production costs will hardly invoke respective changes in production cost in Russia. These changes in relative producer prices reflect differences in industrial process, investments in R&D and, consequently, dynamics of prospective comparative advantages to a greater extent. Due to the non-simultaneous character of these processes in Russia and Germany, the PCA index based on producer prices for a particular industry may change substantially from one year to another. Therefore, these indices must be analyzed for a sequence of years to make a conclusion in regard to the existence of prospective advantages.

Observing the PCA indices of Russia based on manufacturer prices, we can see that Russia has prospects in diversifying its export structure in a wider set of low, medium and high technological industries (see Table 2.6).

Discussion

In the following we will interpret the obtained results of CAs to reveal those industries where the stimulation of comparative advantages may show the most promise.

Manufacturing of electronic equipment is one of the truly infant industries in Russia, since many enterprizes in this sector got bankrupt in the early 1990s, while others were conversioned from the military sector. Since then this industry is characterized as disadvantageous based on the trade statistics. We can also see that the obtained PCA indices based on consumer and producer prices are not consistent: there are prospective advantages in manufacturing of television and radio receivers in terms of producer prices, but in terms of consumer prices there are disadvantages. This is an example when the exchange rate distortion has a significant effect. On the Russian market the major share belongs to foreign-made electronic equipment, and because of the Russian rouble′s undervaluation advantages in this industry based on consumer prices are not observable. In fact, the undervalued rouble gives a competitive edge to Russian companies enabling them to set higher prices and maintain a comparative price level with foreign producers. As the prices in the sector will rise in the course of time, investments in this industry will be relatively more efficient and the prospective comparative advantage of the industry may be realized.

In the clothing sector Russia has some moderate revealed disadvantages. At the same time we need to consider the fact that domestic products are nevertheless competitive on the foreign market (there is also a substantial export of products in this industry). Here we also face the influence of the exchange rate distortion. The overvaluation of prices here is also induced by import tariff rates, which are among the highest for textile and clothing in Russia (Tarr 2007). Therefore, we observe no undervaluation in terms of consumer prices on the market. But based on the producer prices we obtain a much less distorted estimation that points to advantages in this sector. After the Soviet Union collapse many of the existing factories were closed or significantly reorganized, so we can consider this industry as 'quasi-infant'. Considering that Russia has prospective advantages not only in production of clothes but also in textile materials and machinery for textile production, we suggest that this sector can be considered as prospective.

Revealed disadvantages of Russia in the pharmaceutical industry have significantly increased during the last years. At the same time prospective advantages remained more or less stable. There is no doubt that this industry is supported by the Russian government, but considering that this industry has a good basis of related chemical sub-industries in Russia and related research institutes we assume that this industry is prospective. But to realize this potential and to generate new technologies, large investments are essential given that the pharmaceutical industry is one of the most capital-intensive industries in the world (Agrawal 1999).

Russia also has prospective advantages in terms of producer prices in manufacturing machinery, e.g., for food processing and mining. Mining and construction machinery are of high demand on the Russian market as the economy is growing and exploration industries remain prevalent in the Russian economy.

There are also prospective advantages in production of motor vehicles and railway equipment. In these two sectors Russia has revealed disadvantages during the last years. It is widely known that the motor vehicles sector in Russia has a strong public support and its competitive position is poor. This can be a reason for prospective advantages in motor cars in terms of consumer prices as prices on domestic cars are artificially lowered. Thus, we can expect that after economic liberalization this sector will likely be reduced and substituted by international companies and their assembling plants. This process takes place already. In comparison to the motor sector there are some differences with the railway sector. Though companies that produce railway equipment are partly owned by the government, this sector is more competitive on the international market. An evidence for this fact can be observed, if we consider the trade performance of this sector more closely: in sub-industries like self-propelled railway coaches or rail locomotives powered from an external source of electricity there is a positive net export (see Table 2.4). But the industry is lacking in new technologies and investments. Therefore, the high-speed railroads projects in Russia are implemented in a close cooperation with foreign companies. As a result, we consider this sector as 'prospective' for formation of comparative advantages.

2.4 Conclusion

In this paper we demonstrate that the known instruments of comparative advantages theory provide an analysis of 'realized' advantages but do not provide any forecast which industries have a potential to improve their CAs. Nevertheless these forecasts are very important for private companies and public authorities in order to improve their economic strategy. For Russia it is even more important as the economy is still in a period of transition.

To solve this problem we suggest a new approach that forecasts these dynamics based on the relative price differences under the assumption that the trade relations of a particular country are not liberalized yet. Russia, as the biggest economy outside the WTO agreements and other effective free-trade zones like the EU, fits this assumption.

We also show that the Prospective Comparative Advantages mechanism should not be considered independently. In contrary, it needs to be interpreted in conjunction with both revealed comparative advantages and possible exchange rate distortions. It should also be noted that the calculation of the PCA indices based on producer price indices provides less distorted figures on prospective advantages compared to those based on consumer price indices.

We use an example of CEE countries most of which had structural problems in their economies similar to Russia and show that most of the industries in which these countries managed to gain comparative advantages during the last ten years were undervalued in comparison to the EU. An empirical analysis of the PCA forecasting power provides promising results. That supports our claim for the Prospective Comparative Advantages mechanism as a noteworthy forecasting instrument.

Finally, we calculated Russian revealed and prospective comparative advantages, analyzed their dynamics during the last five years, and suggested that the

Russian Federation has prospective advantages in some medium and high technological industries like pharmaceutical industry, electronic equipment, machinery building and railway transport. But to realize this prospective advantages, joint efforts of public and private sectors in optimizing their economic strategy is needed. For the Russian Federation it is also an issue of its innovation policy. Research on the efficiency of stimulating instruments for the sectors remains for future study.

Acknowledgements Valuable comments and suggestions on earlier versions of the paper by Jürgen Meckl, Georg Götz and Matthias Göcke are gratefully acknowledged. The paper also benefitted from recommendations from an anonymous referee of the Central European Journal of Economic Modelling and Econometrics. All remaining shortcomings are our responsability. Financial support from the DAAD is gratefully acknowledged.

2.5 Appendix

Table 2.4: Lafay indices of Russia on different goods.

	Industry	2002	2003	2004	2005	2006
2709	Crude petroleum oils	10,51	11,24	12,3	11,22	13,63
2710	Petroleum oils, not crude	4,32	4,26	4,28	4,75	6,19
2711	Petroleum gases	6,01	5,99	4,79	4,35	6,13
72	Iron and steel	1,93	1,77	2,63	1,43	1,43
44	Wood and articles of wood, charcoal	1,00	0,91	0,87	0,79	0,79
31	Fertilizers	0,64	0,61	0,64	0,58	0,58
74	Copper and articles thereof	0,32	0,29	0,42	0,57	0,57
2701	Coal, briquettes, ovoids	0,32	0,38	0,50	0,42	0,50
71	Pearls, precious stones, metals, coins, etc	1,30	1,21	1,04	0,24	0,24
29	Organic chemicals	0,14	0,14	0,19	0,16	0,16
47	Pulp of wood, fibrous cellulosic material, etc	0,20	0,18	0,15	0,11	0,11
10	Cereals	0,24	0,19	-0,11	0,10	0,10
8401	Fuel element for reactor; Nuclear reactor	0,25	0,24	0,19	0,11	0,10
8411	Turbo-jets, propellers, other gas turbines	-0,03	0,02	-0,03	-0,01	0,04
51	Wool, animal hair, and fabric thereof	-0,02	-0,01	-0,01	0,00	0,00
53	Vegetable textile fibres nes, woven fabric	0,02	0,00	0,00	0,00	0,00
8603	Self-propelled railway/tramway coache	0,01	0,02	0,22	0,05	0,00
8601	Rail locomotives with ext source of energy	0,00	0,00	0,00	0,00	0,00
43	Furskins and artificial fur, manufactures	-0,01	0,00	0,00	-0,02	-0,02
60	Knitted or crocheted fabric	-0,41	-0,75	-0,06	-0,05	-0,05
45	Cork and articles of cork	-0,24	-0,19	-0,13	-0,06	-0,06
42	Articles of leather, harness, travel goods	-0,11	-0,09	-0,07	-0,07	-0,07
1	Live animals	-0,02	-0,02	-0,02	-0,08	-0,08
52	Cotton	-0,22	-0,19	-0,17	-0,11	-0,11
8413	Pumps for liquids; liquid elevators	-0,24	-0,22	-0,23	-0,14	-0,17
86	Railway, tramway locomotives, rolling stock	-0,22	-0,15	-0,30	-0,19	-0,19
61	Articles of apparel, accessories, knit etc	-0,15	-0,15	-0,13	-0,20	-0,20
89	Ships, boats and other floating structures	-0,10	-0,07	0,02	-0,20	-0,20
8429	Self-propelled bulldozer, excavator, etc	-0,14	-0,16	-0,15	-0,15	-0,25
38	Miscellaneous chemical products	-0,44	-0,38	-0,36	-0,30	-0,30
88	Aircraft, spacecraft, and parts thereof	0,67	0,67	0,48	-0,31	-0,31
64	Footwear, gaiters, parts thereof	-0,21	-0,21	-0,16	-0,36	-0,36
94	Furniture, lighting, prefabricated buildings	-0,42	-0,38	-0,35	-0,36	-0,36
28	Inorganic chemicals	-0,84	-0,53	-0,42	-0,43	-0,43
33	Essential oils, perfumes, cosmetics	-0,60	-0,65	-0,63	-0,53	-0,53
48	Paper & paperboard, articles of paper	-0,75	-0,79	-0,70	-0,59	-0,59
73	Articles of iron or steel	-0,74	-0,72	-0,76	-0,86	-0,86
90	Optical, photo, technical, etc apparatus	-0,98	-1,14	-0,98	-1,11	-1,11
2	Meat and edible meat offal	-2,11	-1,62	-1,23	-1,40	-1,40
30	Pharmaceutical products	-1,38	-1,62	-1,53	-1,91	-1,91
85	Electrical, electronic equipment	-2,96	-2,83	-3,26	-4,30	-4,30
87	Vehicles other than railway, tramway	-1,57	-2,47	-3,67	-5,49	-5,49

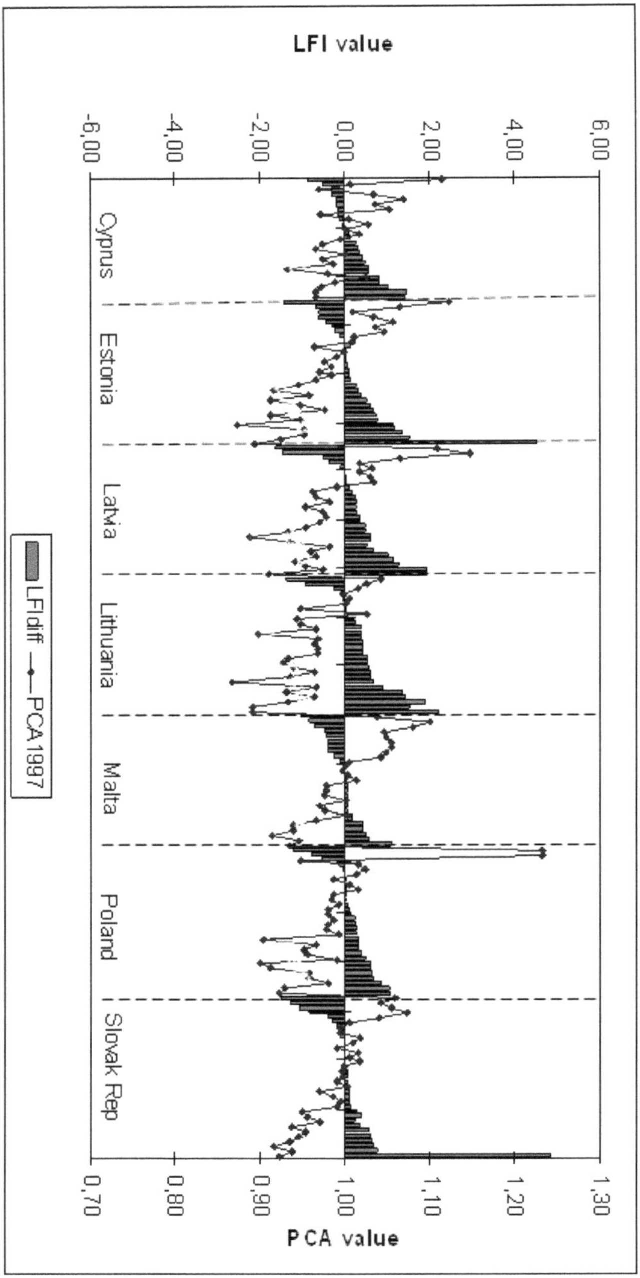

Figure 2.2: Correlation between the LFI index value difference and the PCA index.

Table 2.5: Potential comparative advantages of Russia on different goods based on the consumption price index.

COICOP classification		2003	2004	2005	2006	2007
cp0711	Motor cars	0,98	0,97	0,96	0,97	0,93
cp022	Tobacco	0,92	0,86	0,91	0,97	0,93
cp061	Medical products, appliances and equipment	0,97	0,88	0,94	0,96	0,95
cp0611	Pharmaceutical products	0,96	0,89	0,93	0,96	0,95
cp0311	Clothing materials	0,96	0,96	0,96	1,00	0,95
cp0712_713_714	Motor cycles, bicycles and animal drawn vehicles	0,97	0,96	0,98	1,00	0,96
cp0112	Meat	0,97	1,07	1,05	0,98	0,97
cp0722	Fuels and lubricants for personal transport equipment	1,03	1,13	0,97	0,99	0,97
cp0511	Furniture and furnishings	0,99	0,98	0,98	0,99	0,97
cp10	Education	1,03	1,00	1,02	1,04	0,97
cp031	Clothing	1,01	0,99	1,00	1,02	0,99
cp082	Telephones and telefaxes	0,89	1,06	1,14	1,06	1,01
cp094	Recreational and cultural services	1,08	1,07	1,05	1,05	1,02
cp0612_613	Other medical products, therapeutic appliances and equipment	1,08	0,96	1,07	1,06	1,02
cp112	Accommodation services	1,11	1,05	1,06	1,04	1,04
cp0913	Information processing equipment	1,07	1,03	1,09	1,00	1,05
cp045	Electricity, gas and other fuels	1,07	1,01	0,98	1,01	1,05
cp0111	Bread and cereals	1,16	1,04	0,94	1,02	1,07
cp0911	Equipment for the reception, recording and reproduction of sound and pictures	1,04	1,03	1,04	1,08	1,11

Table 2.6: Potential comparative advantages of Russia on different goods based on the producer's price indices.

NICE classification		2002	2003	2004	2005	2006	2007
dl323	Manufacture of television and radio receivers, other reproducing apparatus	0,81	0,87	0,80	0,97	0,92	0,78
dk2953	Manufacture of machinery for food, beverage and tobacco processing	0,87	0,89	0,93	0,98	0,98	0,80
dm3542	Manufacture of bicycles	0,94	1,02	0,84	0,98	0,96	0,81
dk2911	Manufacture of engines and turbines, except aircraft and vehicle engines	1,00	1,00	0,90	0,94	1,01	0,81
dm341	Manufacture of motor vehicles	0,89	0,99	0,88	0,93	1,02	0,82
db176	Manufacture of knitted and crocheted fabrics	0,96	0,97	0,85	0,92	0,97	0,82
dl331	Manufacture of medical and surgical equipment	0,95	1,02	0,87	0,96	1,00	0,83
dm352	Manufacture of railway, tramway locomotives, rolling stock	0,96	0,94	0,81	1,12	0,99	0,83
dk2954	Manufacture of machinery for textile, apparel and leather production	1,16	1,07	0,82	0,84	0,93	0,84
dm343	Manufacture of parts, accessories for motor vehicles	1,03	0,98	0,86	1,01	0,97	0,85
dk2952	Manufacture of machinery for mining, quarrying and construction	0,84	0,94	0,88	0,94	1,15	0,86
dg2452	Manufacture of perfumes and toilet preparations	1,02	0,97	0,86	0,96	1,23	0,86
db1721	Cotton-type weaving	0,84	1,05	0,87	0,91	0,96	0,87
dn365	Manufacture of games and toys	0,88	1,01	0,80	1,02	1,10	0,87
dn361	Manufacture of furniture	0,93	1,02	0,89	0,94	1,01	0,88
dn3622	Manufacture of jewelery and related articles n.e.c.	0,87	0,99	0,85	0,94	1,06	0,90
dk2922	Manufacture of lifting and handling equipment	0,88	1,04	0,91	0,95	1,03	0,90
dk293	Manufacture of agricultural and forestry machinery	0,86	1,11	0,81	1,01	1,10	0,95
df232	Manufacture of refined petroleum products	0,96	1,00	0,99	0,82	1,02	1,00
dl3002	Manufacture of computers and other information processing equipment	1,01	0,92	0,98	1,09	1,13	1,12

Chapter 3

Heuristic Optimization Methods for Dynamic Panel Data Model Selection

Application on the Russian Innovative Performance

Ivan Savin, Peter Winker
The final publication is available at http://www.springerlink.com

3.1 Introduction

Innovative activity, 'creative destruction', is widely seen as the main factor of economic growth. A vast literature is available on this interrelation (see, e.g., Schumpeter (1943) and Porter (2003)). There is also a large body of empirical research on this issue (see, e.g., Merivate and Pernias (2006)).

However, evidence on the importance of different factors stimulating innovations is mixed. This might to some extent be due to ad hoc or intuitive decisions in the model specification step. Picking up too many regressors increases the variance of the parameter estimates for the model, and taking less regressors than needed results in inconsistent estimates.

In our application, we face the problem of selecting relevant factors explaining the innovative performance of Russian regions based on regional data for the period 1999–2006.

During the last decade several research strategies have been introduced to extract relevant information from large databases. Among these are the general-to-specific approach (PcGets) discussed by Hendry and Krolzig (2005) and its bottom-up alternative (RETINA) analyzed by Perez-Amaral *et al.* (2003). However, in general, there will be no consensus model resulting from the application of these methods. Another option is presented by the least absolute shrinkage and selection operator (Lasso). The Lasso-type estimator is found to be more effective in comparison to the conventional methods, but has an asymptotic bias due to shrinkage (Hsu *et al.* 2007, Savin forthcoming). An alternative model selection approach is based on information criteria (IC), which rank models according to their fitness and a penalty for model complexity (see, e.g., Winker (2000)).

To deal with the problem of model selection, an efficient algorithm is required selecting the model specification with the best value of the IC or at least a good approximation to this optimum. To this end, we compare two selection procedures based on two heuristic optimization approaches: Threshold Accepting and Genetic Algorithms. A new feature of this paper is the application of heuristic model selection methods to a panel dataset with short time series (using the system Generalized Method of Moments estimation method). Because of unobserved heterogeneity and possible endogeneity of regressors we consider a dynamic model specification with fixed effects. Furthermore, additional restrictions on subgroups of regressors derived from economic reasoning are taken into account.

The remainder of the paper proceeds as follows. Section 3.2 introduces both the economic background and the estimation framework for our application. Section 3.3 presents the model selection problem and the heuristic techniques proposed as an effective alternative to the standard procedures. Section 3.4 reports the results of our Monte-Carlo analysis and Section 3.5 presents the results for the real data. Finally, Section 3.6 contains concluding remarks.

3.2 Innovations: Factors and Model Framework

The practical example we analyze is the innovative performance of Russian regions. As no data at the firm level is available, we use regionally aggregated data for the period between 1999 and 2006 from the 'Regions of Russia: Social-economic indicators' database (Rosstat). The quality of the data is not perfect. Hence, our conclusions should be considered rather suggestive than irrevocable. Our main goal is to introduce the new method of model selection in dynamic panel data models.

Analyzing the data, Russian regions can be considered as 'potential innovative clusters' (Porter 2003). Among the main actors of any cluster are companies, financial and educational institutions, public authorities and specific cluster organizations specialized in transferring knowledge and providing further services. But these clusters are potential in the sense that not in every region, and not in the frame of a whole region effective clustering occurs. For further details on the theory of innovative clusters see Sölvell (2008).

There are numerous studies devoted to specific factors triggering innovative activity (see, e.g., Funk (2006)). Nevertheless, as far as we know, no generally accepted model encompassing all the factors of interest is available.

Our main indicator of innovative activity is the value of innovative output of organizations[1] in a region. This is in line with Rosenberg *et al.* (1992) who argue that the better measure of innovative success is not technology itself, but its market success. The data of Rosstat do not allow to distinguish different types of innovations, including new products and services, which are not offered or produced in the country, and processes, which increase efficiency of the production. As a result we also consider a larger number of organizations related to innovative processes by implementing established technologies.

3.2.1 Advanced Hypotheses and Data Description

To identify the driving factors of innovations, we split the database in eight groups of variables according to the hypotheses tested in this study. The full list of variables as well as the list of regions included in the dataset can be found in Tables 3.5 and 3.6 in Appendix 3.7.1.

1. Product market competition. There is a long discussion in the theory of industrial organization on whether competitive pressure induces or reduces innovative output of companies. Firstly, according to Schumpeter (1943), there is a negative correlation between innovative activity, 'creative destruction', and competitive pressure as profits become too small to implement innovations. On contrary, Blundell *et al.* (1999) empirically confirmed that competition enhances R&D activity in order to gain an advantage towards competitors. During the last years, the idea of an inverse 'U-curve' dependence of innovative activities on the degree of competition has become popular (Bucci and Parello 2009). Previous research based on survey data for Russia confirmed the existence of an inverted 'U-relationship' (Rautava (2005)).

Dealing with the regionally aggregated data of Rosstat, we can use neither standard indices of the extent of market competition like the Lerner Index nor the number of competitors as a proxy measure. We can only approximate the number and percentage share of companies that produce innovative output, conduct R&D, apply and register patents and implement advanced technologies in

[1]Organizations according to the Russian Civil Code are public and private companies as well as scientific institutes and high-schools.

their production in a particular region. This substitution has certain disadvantages, e.g., it does not differentiate between industries, where innovative firms actually act, but is included in order to compare results.

2. Scale of production. An intensive discussion can also be found on the role of small and medium-sized companies (SMEs) in developing innovative products. From one point of view, big companies benefit from economies of scale and scope and are rather expected to be more innovative. But at the same time, SMEs are more willing to undertake risks and are more flexible to react to changes in consumer preferences (Merivate and Pernias 2006).

3. Form of ownership. We distinguish between three types of ownership. In the case of public property, there are few incentives for managers to run a business in the best way; it is believed that international corporations carry out most of their research in their headquarters; the local private capital is widely seen as the most efficient owner of innovative enterprizes (Jefferson *et al.* 2003). In our model we are interested in revealing whether public ownership has an impact on the innovative activity and in testing possible correlation between foreign investments and innovative output.

4. Economic performance. There are two controversial opinions in regard to dependence of innovative activities on the economic performance of companies. On one hand, companies that face financial difficulties try to diversify their activities by implementing innovations (Funk 2006). On the other hand, there is empirical evidence that companies with stable profits in previous years adopt and implement innovations more actively (Cainelli *et al.* 2006), which might be a result of financial constraints (Winker 1999).

5. Infrastructure. It is argued that the actual level of infrastructure has an important impact on innovative performance. By improving infrastructure, significant reductions in transaction costs and, hence, an improvement in the market efficiency in general may be obtained. All these factors induce restructuring processes in companies and introduction of new products. Among infrastructure factors accelerating innovations the most important are transport, telecommunication and banking services (Cainelli *et al.* 2006).

6. and 7. Knowledge spillovers. Inter-regional spillovers describe potential benefits from cooperation with other innovative clusters. Knowledge diffusion has an important role in fostering innovative performance, especially in developing countries. This study concentrates on two aspects.

First, the regional trade activity and ability to absorb new knowledge as factors, inducing innovations (MacGarvie 2001). Considering factors that improve this transfer, the level of education and infrastructure are most important. We also test the share of export and import with the Commonwealth of Independent States and other countries and the trade agreements on technologies.

Second, we test R&D activity in neighbor regions as well as their education level as stimulating factors for innovative activity in a particular region. There are numerous methods to determine one's spatial neighbors, for instance, contiguity matrices and distance-decay-functions (Klotz 1997). Another alternative based on the Haynes model being able to estimate innovation diffusion process is presented by Shinohara and Okuda (2010). However, the latter approach can be only applied to particular innovations and requires microdata which we do not have at our disposal. In our study we define only direct neighbors by land as neighbor regions and calculate respective variables as arithmetic means of those in neighbor regions.

8. Control variables. We also aim to test some general hypothesis on regional socio-economic characteristics stimulating innovations. It is of particular interest to investigate whether innovations are higher in economically strong regions with large GRPs and budget revenues. Among variables tested are long term assets value, shares of urban population, unemployment and criminal activity.

Innovations can be explained by the economic performance of companies, investments in infrastructure, and product market competition. However, they also induce improvements in economic performance, further investments in infrastructure and increase market competition (Cainelli *et al.* 2006). Therefore, we face the problem of a potential simultaneity bias. In order to tackle this problem, we use the instrumental variable approach in the context of a dynamic model specification.

3.2.2 Model Specification

In spite of the huge number of models proposed to explain the innovation process, there is no generally accepted model available encompassing all the factors of interest. Reviewing a great diversity of models, Forrest (1991) suggested some essential characteristics for a comprehensive model. First of all, the model should be nonlinear capturing an interrelationship of various stages of innovative activity. Then, it should include 'identifiable' inputs and outputs. In addition, the generalized model must incorporate external effects, e.g., market competition and indicators of socio-economic environment. After all, the model should take into account the possible heterogeneity of regions.

In order to set up such a generalized model, we consider the modified Cobb-Douglas Knowledge Production Function (KPF) (Crescenzi *et al.* 2007, p. 170). Transforming it into a log-linear form approximating the initial model with arguments according to the hypotheses stated above we obtain:

$$lnY_i = \alpha + \beta_1 lnPMC_i + \beta_2 lnSME_i + \beta_3 lnFO_i + \beta_4 lnEP_i \\ + \beta_5 lnInfra_i + \beta_6 lnSAbs_i + \beta_7 lnSN_i + \beta_8 lnCV_i + u_i, \qquad (3.1)$$

where:

Y_i	innovative output of region i;
α	a constant;
PMC_i	indicators of market competition in region i;
SME_i	indicators of SME's activity in region i;
FO_i	proxies for ownership structure of companies;
EP_i	proxies for companies' economic performance;
$Infra_i$	proxies for regional infrastructure development;
$SAbs_i$	regional socio-economic characteristics, which may improve ability to absorb new knowledge;
SN_i	socio-economic characteristics in neighbor regions;
CV_i	further control variables.

Moving from the KPF to (3.1), we approximate the stock of initial knowledge by several proxies of regional socio-economic characteristics ($SAbs_i$), e.g., the number of doctoral students and employees in R&D departments. The number of patent applications (Crescenzi *et al.* 2007), the number of patents granted and the regional innovative output in the previous period (for the dynamic model specification) are also used as proxies for the stock of initial knowledge. Regional R&D activity is proxied by some market competition indicators (PMC_i), e.g., the internal R&D costs.

Due to the logarithmic transformation in (3.1), β is an elasticity: the percent change in Y as a function of the percent change in the respective variable. As the transformation can be applied only to strictly positive data, some variables are expressed as percentage shares or average values. For these variables, β is a rate of proportional change in Y per unit change in the respective regressor.

Gathering dependent and explanatory variables in vector y and matrix $x' = (x_1, ..., x_k)$, respectively, model (3.1) can be written as:

$$y_{it} = \alpha\iota_{NT} + x'_{it}\beta + u_{it}, \; i = 1, ..., N \; t = 1, , ..., T, \qquad (3.2)$$

where α is a scalar, ι_{NT} stands for an $NT \times 1$ vector of ones, x' is an $NT \times k$ matrix of k regressors and their values for N regions and T periods, β is a $k \times 1$ vector of their coefficients and u is an $NT \times 1$ vector of residuals. Here we use the panel data subscripts i for regions and t for time.

Application of the Hausman test (Hausman and Taylor 1981) to preliminary estimates of (3.2) indicates that regional fixed effects (μ_i) should be taken into account. To deal with these fixed effects we first specify a static model (3.3), where Z' stands for the $NT \times N$ matrix of regional dummies and ν_{it} denotes the remainder disturbance:

$$y_{it} = \alpha\iota_{NT} + x'_{it}\beta + Z'_i\mu + \nu_{it}. \tag{3.3}$$

Transforming the data in (3.3) into deviations from individual means, the least squares dummy variables (within) estimation can be performed.

Furthermore, due to potential endogeneity of some of our regressors, we consider a dynamic model suggested by Blundell and Bond (1998) with instruments in levels suitable for panel data with short time dimensions:

$$y_{it} = \delta y_{i,t-1} + x'_{it}\beta + Z'_i\mu + \nu_{it}. \tag{3.4}$$

The GMM method for dynamic panel data models with no strictly exogenous variables was developed by Arellano and Bond (1991), introducing some basic restrictions on the model, e.g., no serial correlation, and using values of y_{it} and x'_{it} lagged two periods or more as instrumental variables in equations with first-differences. Later, it was shown that this estimator is weak and biased (see, e.g., Okui (2009)). The system GMM procedure introduced by Blundell and Bond (1998) appears superior, as it adds lagged differences of y_{it} and x'_{it} as additional instruments in order to improve the efficiency for short time-series samples.

We consider two scenarios, where the explanatory variables x' are considered either as endogenous or predetermined. Depending on which of these assumptions is maintained, different numbers of lags and lagged differences as instruments are used in the system GMM estimation. To test the validity of the instruments we apply the Sargan test (ST). The details of the system GMM estimation procedure are presented in Appendix 3.7.2.[2]

3.3 Model Selection Procedure

3.3.1 Optimization Problem

Let us first clarify the basic approach to the optimization problem. Consider the following regression function:

[2]We also apply the system GMM estimation procedure implemented in Stata 10 as a reference to compare our results.

$$y_{it} = \alpha \iota_{NT} + (x_{it}^{opt})' \beta + u_{it}, \tag{3.5}$$

where $x' = (x_1, ..., x_k)$ is a k-dimensional vector of variables with $(x^{opt})' = x'\omega^{opt}$ being the subset of all possible regressors we seek to identify. This might be the 'true' model in a Monte-Carlo simulation setting or an optimal approximation to the unknown real data generating process. A $k \times 1$ vector ω consisting of ones and zeros specifies which variables are, respectively, included in the model and which are not. Thus, for each potential regressor in our dataset we assign a binary value in the vector ω. In the following we will refer to the sum of these values as h.

To select a model IC are applied, which rank alternative models according to their fitness, while taking into account a penalty for model complexity. Over the last years IC became a standard instrument in model selection problems ranging from lag order selection in multivariate linear (VAR and VEC) and nonlinear (MS-VAR) autoregression models to selection between rival nonnested models (Winker 2000, Gatu *et al.* 2008, Kaucic 2009).

In this study we implement the Akaike's IC (AIC), the Bayesian IC (BIC) and the Hannan-Quinn IC (HQIC). All these criteria have a similar structure:

$$\text{IC} = \ln(\widehat{\sigma}^2) + f(h, NT), \tag{3.6}$$

where $\widehat{\sigma}^2$ is the maximum likelihood estimation of the residual sum of squares, which should be as small as possible for a high model fitness. The second term is a penalty for the number of included parameters (h) in relation to the number of observations (NT).[3] In particular, $2h/NT$, $h\ln(NT)/NT$ and $2h\ln(\ln(NT))/NT$ are the AIC, BIC and HQIC penalties. Thus, the smallest IC corresponds to the preferable model specification.

Imposing some weak assumptions on the model space ($x_{i,t}$ and $\varepsilon_{i,t}$) according to the results of Sin and White (1996) it can be shown that the vector ω that minimizes the IC converges to ω^{opt} with probability close to 1 as $NT \to \infty$. But for this to be true, it is essential that the penalty term $f(h, NT) \to \infty$ and $f(h, NT)/NT \to 0$ as $NT \to \infty$. In this sense, BIC and HQIC are consistent, while AIC is inconsistent.

In addition to the penalty for model complexity, we impose the constraint that at least one regressor from each group of variables specified in (3.1) is included. This constraint is enforced by imposing an additional multiplicative penalty (p_j) to the objective function (3.6) in an optimization procedure. The penalty increases over the iterations of the optimization algorithm to make sure

[3] Analyzing the panel dataset of Rosstat over 75 regions (N) and eight years period (T) we have a sample size of 600 observations (NT) per variable. (which has to be minimized as well)

that eventually the constraint is satisfied. This constraint is optional, as we might realize that no statistically significant variable is present for a particular group. Furthermore, for the dynamic specification of the model the objective function is multiplicated by an additional penalty term ($penST$) derived from the results of the Sargan test to ensure that only orthogonal instruments are included in the final model estimation.[4] Similar to the 'group constraint', this penalty grows during the optimization process ensuring that the resulting solution passes the Sargan test:

$$\text{IC} = (\ln(\widehat{\mathrm{v}}^2) + f(h, NT))\left(1 + \sum_{j=1}^{8} p_j\right)(1 + penST), \tag{3.7}$$

where j stands for a group of variables and $\widehat{\mathrm{v}}$ denotes residuals from the two-step estimator (see Appendix 3.7.2) and

$$penST = \begin{cases} 0 & \text{if } ST_{prob} > 0.1 \\ 1/ST_{prob} & \text{otherwise}, \end{cases} \tag{3.8}$$

with ST_{prob} being the probability value of the Sargan test.

3.3.2 Heuristic Algorithms

Quality and precision of econometric estimation is crucially dependent on detecting the global optimum of any objective function. Breiman (2001) demonstrates the so called 'Rashomon Effect', where different model specifications with very similar IC values provide different conclusions. Minimizing (3.7) is not as simple as it might seem at first sight. In fact, the search space of candidate models is discrete. The full enumeration of all solutions is only feasible for a small k. In our empirical problem the selection is made out of $k = 80$ variables resulting in 2^{80} potential sub-models. Therefore, the full enumeration is infeasible even using efficient algorithms (Gatu *et al.* 2008).

In the last two decades, new nature-inspired optimization methods have become available. For an overview of these optimization techniques see Winker (2001) and Gilli and Winker (2009). In the following we describe the two heuristic methods implemented, the Threshold Accepting (TA) and the Genetic Algorithms (GA). Both methods have been used previously in the context of model selection (Winker 2000, Kapetanios 2007), but no evidence on their relative performance was provided. This study is considered to cover the gap.

[4]This is similar to the approach presented in Kaucic (2009), where an IC-based model selection via heuristics is penalized in order to control the sensitiveness to outliers.

Threshold Accepting

The TA algorithm is a refinement of classical local search procedures. In contrast to a local search, where a new solution is accepted only if an improvement is realized, TA also accepts uphill moves as long as they do not exceed a given threshold value τ. A pseudocode of the TA implementation can be found in Algorithm 1.

Algorithm 1 Pseudocode for Threshold Accepting.

Generate at random a solution ω^0, initialize I_{max} and τ
for $I = 1$ to I_{max} **do**
    Generate at random neighbor $\omega^1 \in \mathcal{N}(\omega^0)$
    if $f(\omega^0) - f(\omega^1) < \tau$ **then**
        $\omega^0 = \omega^1$
    end if
    Reduce τ
end for

In TA we generate an initial solution ω^0 as a vector of k binary components corresponding to potential predictors. A fixed number of variables (here it is 2) is included in each group and they are randomly distributed across the vector.

We generate a new solution ω^1 by exchanging two randomly chosen components with components located in a close neighborhood,[5] in particular in the radius of three vector components. We choose the 'neighbor' by means of the uniform random distribution and if it has the same value as the first component, than its value is changed to an opposite one. The same is true if the 'neighbor' turns out to be the initially chosen component. For more details, e.g., regarding the generation of the threshold sequences, see Savin and Winker (2012).

As TA is a stochastic process it may find the best possible solution and then loose it. To avoid this, the best found solution is saved.

Genetic Algorithms

Unlike TA, GA are population based heuristics that operate on a set of solutions. GA investigate the search space in many directions simultaneously and perform certain jumps in the search space by crossover and mutation of solutions in the population. Hence, the probability of getting stuck into a local optimum is reduced.

The members in the GA population (chromosomes) are represented as bit strings, in which each position (gene) is either 1 or 0. In each generation GA

[5]Changing the value of only one element reduces the speed of convergence and, e.g., of four elements causes a larger variance of results.

replace parts of a population with new chromosomes (children) aimed to represent better solutions for a particular problem. For optimal model selection we implement the GA pseudocode described in Algorithm 2.

Algorithm 2 Pseudocode for Genetic Algorithms.

Generate initial population K of solutions, initialize G_{max} and C
for $g = 1$ to G_{max} **do**
  Sort chromosomes in K
  Select $K' \subset K$ (parents), select $K^* \subset K$ (elitist)
  initialize $K'' = \emptyset$ (set of children)
  for $c = 1$ to C **do**
    Select individuals $x^{parent1}$ and $x^{parent2}$ at random from K'
    Apply cross-over to $x^{parent1}$ and $x^{parent2}$ to produce x^{child}
    $K'' = K'' \cup x^{child}$
  end for
  $K = (K', K'', K^*)$
  Mutate $K \setminus K^*$ at 8 random points
end for

K is a matrix of p initial solutions. We use $p = 500$ considering this number to be large enough to screen the search space in different directions and at the same time small enough to allow for effective sorting and selection of the best solutions.[6] As in TA, chromosomes in K are generated with a fixed number of included variables. Thereafter, the population is sorted in an ascending order according to the objective function (3.7). Then, the 50% of the chromosomes with the best target values (parents, K') are transferred to the new population. Based on K' we construct new chromosomes (children) by crossing them over. First, we select 100 pairs of parents at random with an equal probability to be selected and generate 200 children. Then, the 40 parents with the best objective values generate 40 more children. The ten last children are generated from the best ten solutions (elitists, K^*) by changing them at one random gene.

In this study we compare the single-point and the uniform crossover mechanisms. In the single-point crossover two parents are split at a random gene and their parts are combined forming two children. In the uniform crossover parents may be split at each gene. With probability P_0 we swap genes from two parents in a child. The uniform crossover can be presented as generating a mask of zeros and ones (Figure 3.1), indicating from which parent a gene has to be taken. We set $P_0 = 0.5$ resulting in equal probability for both parents.

Testing both mechanisms for this particular problem based on 500 repeated and independent Monte-Carlo restarts we see that the uniform mechanism provides high quality solutions more reliably. The 95th percentiles of the results for the uniform crossover converge to the minimum value found in all replications

[6] We also tested populations of 100, 300 and 1000 solutions and found that the population of 500 solutions is more effective in terms of both CPU time and solution quality.

$$
\begin{array}{rl}
x^{parent1} & = \begin{pmatrix} 1 & 1 & 0 & 1 & 0 & 1 & 0 & \dots & 1 \end{pmatrix}_{1\times k} \\
x^{parent2} & = \begin{pmatrix} \mathit{1} & \mathit{0} & \mathit{1} & \mathit{0} & \mathit{1} & \mathit{1} & \mathit{0} & \dots & \mathit{1} \end{pmatrix}_{1\times k} \\
\hline
mask_1 & = \begin{pmatrix} 0 & 1 & 1 & 0 & 0 & 0 & 1 & \dots & 1 \end{pmatrix}_{1\times k} \\
mask_2 & = \begin{pmatrix} 1 & 0 & 0 & 1 & 1 & 1 & 0 & \dots & 0 \end{pmatrix}_{1\times k} \\
\hline
x^{child1} & = \begin{pmatrix} \mathit{1} & 1 & 0 & \mathit{0} & \mathit{1} & \mathit{1} & 0 & \dots & 1 \end{pmatrix}_{1\times k} \\
x^{child2} & = \begin{pmatrix} 1 & \mathit{0} & \mathit{1} & 1 & 0 & 1 & \mathit{0} & \dots & \mathit{1} \end{pmatrix}_{1\times k}
\end{array}
$$

Figure 3.1: The uniform crossover mechanism.

(see Figure 3.2), whereas for the single-point crossover a difference between the 5th and the 95th percentiles persists.[7] More information on performance analysis of the uniform crossover method based on binary function optimization problems can be found in Fogel (2006).

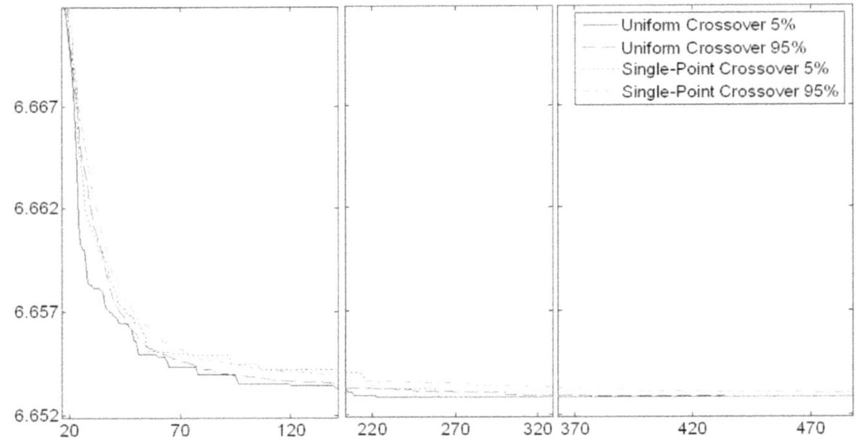

Figure 3.2: Results of the single-point and uniform crossover.

The uniform crossover might be criticized for destroying superior chromosome structures. We avoid this problem by preserving elitist solutions.

After a new population is formed, mutation is applied at eight random genes with a probability of 50%.[8] Mutation is applied to the whole new population K except for the 10 elitist solutions and the 10 children generated from the

[7]In Figure 3.2, all objective function values accepted by the GA lie within the interval between 6.65 and 6.72.

[8]By reducing the number of genes or the probability of mutation, the speed of convergence decreases, while increasing both values increases the risk to miss high quality solutions.

elitist solutions by mutation. This procedure is repeated for a given number of generations G_{max}.

Both heuristics implemented have a series of tuning parameters that have to be set adequately for a given problem. Comparing TA and GA one could conclude that GA are a more sophisticated method since they include more parameters to be tuned. However, this makes the algorithm more challenging for an unexperienced user. Meanwhile, there exist guidelines on how to tune heuristic algorithms (see Gilli and Winker (2009)).

3.4 Monte-Carlo Study

3.4.1 Data Generating Process

In order to assess the performance of TA and GA with an objective function as described in equations (3.6) and (3.7) we generate artificial data based on the panel dataset of Rosstat. First, a set of regressors (x'_{MC}) is randomly drawn from the database. Then, regression coefficients (β_{MC}) are estimated based on the dependent variable (y). Finally, a new dependent variable (y_{MC}) is generated using β_{MC} and adding an i.i.d. error term:

$$y_{MC} = x'_{MC}\beta_{MC} + \varepsilon, \quad \varepsilon \sim N(0, \sigma_\varepsilon^2), \tag{3.9}$$

where σ_ε^2 is the variance of the residuals.

Using a Data Generating Process (DGP) mimicking the empirical data, i.e., also with a cross-section dimension of 75 and eight time periods, we expect that the performance of the heuristics and the IC estimated based on this Monte-Carlo study is a good approximation for our real data problem.

3.4.2 Simulation Results

Table 3.1 presents results of TA and GA computational performance. The two algorithms are compared in terms of mean and minimum objective function values and standard deviations. The descriptive results are obtained based on 10 restarts for each algorithm. The number of iterations for TA is taken equal to the number of chromosomes times the number of generations for GA resulting in the same number of function evaluations for both algorithms.

For TA with an increasing number of iterations we observe that all three statistics decrease for all IC (see Table 3.1). For GA we obtain similar results, though the improvement is moderate. Comparing the results we see that GA

Table 3.1: Performance of the algorithms for different computing times.

	Method					
	Threshold Accepting			Genetic Algorithms		
	AIC	BIC	HQIC	AIC	BIC	HQIC
	50 000 iterations			100 generations		
mean	6.5890	6.7322	6.6663	6.5890	6.7272	6.6556
std	[0.0060]	[0.0086]	[0.0093]	[0.0024]	[0.0021]	[0.0011]
min	(6.5877)	(6.7192)	(6.6690)	(6.5844)	(6.7251)	(6.6544)
	250 000 iterations			500 generations		
mean	6.5887	6.7264	6.6646	6.5848	6.7158	6.6528
std	[0.0041]	[0.0053]	[0.0056]	[0.0013]	[0.0017]	[0.0000]
min	(6.5863)	(6.7173)	(6.6532)	(6.5837)	(6.7142)	(6.6528)
	1 000 000 iterations			2000 generations		
mean	6.5876	6.7163	6.6536	6.5843	6.7142	6.6528
std	[0.0018]	[0.0008]	[0.0004]	[0.0011]	[0.0000]	[0.0000]
min	(6.5854)	(6.7153)	(6.6529)	(6.5837)	(6.7142)	(6.6528)

are able to find a good solution already with a relatively small number of generations (in particular, with 500 generations). Increasing the computation up to 2 000 generations mainly reduces the variance of the results. Further increases of CPU time (e.g., up to 5 000 000 iterations for TA and 10 000 generations for GA) do not improve the results significantly. We conclude that GA appear superior to TA for this particular problem. First, GA are about 20 percent faster than TA for a comparable number of iterations.[9] Second, GA provides us with smaller standard deviations and slightly better objective function values (in terms of both mean and minimum values).

Now, to test the performance of our algorithms with an objective function of type (3.7) in detecting 'true' variables we run the algorithms for ten different artificial sets of data with eight 'true' variables in each dataset. The simulation results are compared using the True Positive Rate (TPR) and the False Positive Rate (FPR)[10] in the upper panel of Table 3.2.

As expected the objective function based on the Akaike criterion accepts too many 'false' variables. Accepting on average 19-20 variables, AIC correctly identifies approximately 4 variables. BIC and HQIC significantly outperform AIC, accepting less false variables (on average 8 and 11 regressors in total, re-

[9]Both algorithms are implemented using Matlab 7.7 on a Pentium IV 2.67 GHz. The CPU time for 2 000 generations of GA is $\approx$ 700s, while TA with 1 000 000 iterations requires 850s. The small difference is due to the more complex generation of neighbors for TA.

[10]TPR is the percentage of 'true' regressors from all regressors selected. FPR is the percentage of selected 'false' regressors among all 'false' regressors.

Table 3.2: Performance of the algorithms with group penalty.

	Method					
	Threshold Accepting			Genetic Algorithms		
	AIC	BIC	HQIC	AIC	BIC	HQIC
	1 000 000 iterations			2000 generations		
	8 years period					
TPR	23%	43%	37%	30%	49%	38%
FPR	21%	8%	10%	16%	6%	9%
	20 years period					
TPR	30%	55%	48%	35%	58%	49%
FPR	19%	5%	8%	15%	4%	7%

spectively). As it is clear from Table 3.2 BIC is the most efficient IC in declining 'false' variables, while HQIC regularly allows for more 'incorrect' variables in the final solution.

We believe that the main reason for the limited efficiency of all IC in Table 3.2 is the relatively small sample size: 75 regional observations for a period of eight years are not sufficient for the IC to identify the 'true' model. In fact, using an optimization technique we always find an IC-value smaller than the one corresponding to the 'true' model.

In order to test the performance of the IC for a larger sample size, we artificially increase the time-series dimension: we select 'blocks' of length of three years from the dataset at random points (from 1 to 8) and add them to the current dataset. Selecting the blocks at the same point for all variables, we produce a new dataset of 20 years. Although the dynamic structure of these artificial data differs from the one of the real data, we might still use them to test the performance of the selection criteria for larger sample sizes. Results of these experiments are presented in the lower panel of Table 3.2. It is obvious that the performance for all IC is significantly improved. Unfortunately, actual data for a longer period is not available for Russia.

Considering again the original structure with eight periods, we also analyze the performance of the IC when no penalties on the number of groups included are introduced (see Table 3.3). Thus, in the Monte-Carlo experiment with eight 'true' variables, one from each group, BIC selects four variables on average with three to four of them correct and HQIC selects eight variables on average with four to five correct, accepting more false regressors. Increasing the number of a priori 'true' variables the algorithms marginally increase the number of selected variables, improving their performance in terms of both: the TPR and FPR. For example, for 37 'true' variables (five variables in each group except of scale of production) HQIC selects 14 variables with 12 correct. However, it is obvious that for all IC this effect is accompanied with an increase of the number of true variables not included in the final model. This is also due to the finite sample size, where the asymptotic properties of the IC come into force to a limited extent. In Tables 3.2–3.3 it is also clear that GA have a tendency to reject more 'false' variables, which might be due to the smaller variance in the objective function values obtained. These facts should be taken into account when interpreting the results in Section 5.

Table 3.3: Performance of the algorithms with no group penalty.

	Method					
	Threshold Accepting			Genetic Algorithms		
	AIC	BIC	HQIC	AIC	BIC	HQIC
	1 000 000 iterations			2000 generations		
	8 'true' variables					
TPR	25%	84%	48%	39%	89%	54%
FPR	19%	1%	6%	14%	1%	5%
	16 'true' variables					
TPR	40%	85%	61%	44%	93%	66%
FPR	18%	1%	6%	13%	1%	3%
	37 'true' variables					
TPR	61%	87%	75%	66%	96%	83%
FPR	20%	2%	7%	19%	1%	6%

3.5 Empirical Results on the Example of Russia

Based on the Monte-Carlo simulation results, we apply the superior GA algorithm on the database of Russian regions in order to specify a log-linear model of the form given in (3.1).[11] We use the objective function (3.6)-(3.7) with AIC, BIC and HQIC with and without the penalty term on groups of variables included. The empirical results are obtained by running GA 10 times with 2 000 generations for each IC. We only present the model specifications related to the smallest objective function values.

In the following we concentrate on results obtained for the dynamic model specification (3.4) using the system GMM method (see Table 3.4) as the static

[11]Basically, very similar (and even the same) models are obtained via TA, although they cost some more computational time and exhibit a higher variance (results on the comparison of the two algorithms via the Hamming distance can be obtained on request).

model estimator might be significantly biased due to endogeneity of regressors. Results on the static model (3.3) can be obtained on request.

An argument supporting the relevance of the obtained results is the fact that the set of regressors included is relatively stable for both assumptions on predetermined and endogenous variables. Another evidence for this is the fact that statistically significant[12] variables are included with and without the 'group penalty', while insignificant variables are dropped in specifications without this constraint (see results on TPR in Tables 3.2-3.3).

According to the results in Table 3.4 the number of granted patents and the number of advanced production technologies used have, ceteris paribus, a considerable positive effect on the regional innovative output. This result does not allow us to make any statement about the state of competition and its impact on the innovative activity. However, this finding provides support to the strategy of catching-up for Russia (Polterovich 2009). In fact, in 2006 only about 7% of the advanced technologies in the country were 'new' not only on the national, but on the international level. Hence, by implementing foreign state-of-the-art technologies Russia accumulates knowledge, reduces its 'distance to frontier' (Acemoglu *et al.* 2003) and increases own innovative potential. In contrast to generation of novel technologies, this strategy is seen to be more realistic for a transition country since it allows it to use the 'advantage of backwardness' (Polterovich 2009).

One way to promote international knowledge diffusion is to stimulate big companies. In our study no significant indicator on scale of production is identified. However, according to theoretical and empirical studies (see Acemoglu *et al.* (2003) and Polterovich (2009)) vertically integrated companies have a key role in imitating new technologies, while the role of SMEs increases as the country approaches the 'frontier'.[13] Since adoption of new technologies is expensive, one can provide loan guarantees, improve protection of intellectual property rights and secure public contracts (see Polterovich (2009)) for respective companies.

According to the strategy of catching-up foreign direct investments (FDI) have a leading role in transferring new knowledge. In our study a positive partial effect of FDI is identified. This is one of the few robust findings of this effect

[12]The significance test is based on asymptotic standard errors for the one-step estimator, which are seen to be more reliable if the residuals obtained from the estimator are heteroscedastic (see Arellano and Bond (1991) and Blundell and Bond (1998)).

[13]This also effectively explains our empirical finding. Due to the fact that in the majority of industries Russia lags behind the technological frontier (with an exception of defense and aero-space ones, where SMEs are atypical), we do not detect any significant SME's effect. Big enterprizes, for their part, facing high risks and uncertainty in Russia (Polterovich 2009) are concentrated in low technological industries and short-term objectives.

for the Russian economy (Tytell and Yudaeva 2006).[14] However, the effect of FDI is fairly low. This can be explained by two reasons. First, in our dataset we have a broader definition of innovation used by Rosstat (see Section 3.2). Second, there is a low FDI inflow itself in relation to GDP in Russia (Tytell and Yudaeva 2006). To improve the efficiency of FDI, certain requirements can be introduced ensuring that new technologies and managerial techniques will be adopted in the host country. For instance, similarly to China one can enforce foreign companies to employ nationals, including top positions (for more details see Polterovich (2009)).

In conjunction to the catching-up strategy it is useful to consider the results on the hypotheses of knowledge spillovers. On the knowledge diffusion within a region only the share of graduates from other public educational institutions (vocational and technical schools, colleges) is selected and is found to be significant. In this case, the regression coefficient is negative, demonstrating that the share of graduates has, ceteris paribus, a negative impact on the innovative success. Taken in relation to the total regional population this indicator varies from its value around 1%, having a marginal impact on the innovative output. Nevertheless, the evidence is not obvious as the graduates of these schools are expected to enhance technical innovations.

According to Vinogradov (2008) during the last 20 years this integral part of education system in Russia has had a tough time. Since these institutions used to work in the planned economy and in a close cooperation with enterprizes (mainly defensive-industrial complex), after the Soviet collapse they faced a drastic financial deficit and lost their business partners. The problem is complicated by a selection bias: since the reputation of these institutions has significantly fallen, potential students prefer a university education. Hence, nowadays these schools have a limited ability to prepare high-quality specialists exhibiting the negative impact on the innovative output.

To improve the situation a series of measures both from the side of public authorities and private sector is required. Industrial companies being the main consumer of the education services can support professional schools on a material-technical basis, provide internship opportunities and employ their graduates; vocational schools can in their turn adjust their programs to the needs of respective enterprizes (more details on the cooperation see, e.g., in Vinogradov (2008)). Public authorities need to take a coordinating role in this process providing economic incentives for both sides of the partnership. A similar strategy can be also implemented in the higher education enhancing cooperation between industry and scientific institutions.

---

[14]Note that for the static model FDI were not selected, and in the dynamic one it is included by all IC with the exception of BIC that generally rejects more true variables (see Section 3.4.2).

Table 3.4: System GMM estimation results.

	Endogenous regressors						Predetermined regressors					
	With group penalty			No group penalty			With group penalty			No group penalty		
Regressors	AIC	BIC	HQIC	AIC	BIC	HQIC	AIC	BIC	HQIC	AIC	BIC	HQIC
Lagged dependent variable	0.18***	0.19***	0.15***	0.19***	0.29***	0.24***	0.20***	0.24***	0.17***	0.19***	0.25***	0.17***
shInnovOrg	0.04*	-	-	0.03*	-	–	0.02	-	-	-	-	-
lnGrantPatent	0.24***	-	0.21***	0.26*	0.26**	0.24**	0.22**	0.22*	0.17*	0.23*	0.22*	0.18*
lnAdvProdTech	0.21***	0.43***	0.30***	0.21**	0.29**	0.23**	0.29***	0.34***	0.29***	0.32***	0.25***	0.33**
shSMEoutput	0.004	0.004	0.001	-	-	-	0.002	0.004	0.004	-	-	-
lnFDI	0.05**	0.05*	0.05*	0.04**	-	0.05**	0.03*	-	0.03*	0.04**	-	0.05**
shPriInvFixCap	0.01	-	-	0.01	-	0.01*	0.01	0.01	0.01	0.01	-	0.01**
shAggrNetPinGRP	0.02***	0.02***	0.02***	0.02***	0.03***	0.03***	0.02***	0.01*	0.02**	0.02***	0.02***	0.01*
avNetProfit	0.06	-	0.09	-	-	–	-	-	0.04	-	-	-
lnAggrCredDinRu	0.08	0.09	0.08	0.02	-	0.01	0.04	-	-	0.05	-	-
lnRailRoadDen	0.14**	0.15***	0.13***	0.11*	0.17**	0.15***	0.02	-	-	-	-	-
lnTurnRailRoad	0.01	-	-	-	-	–	-	-	-	-	-	-
lnTurnMotorRoad	-	-	-	-	-	–	0.21	0.24	0.34***	0.19***	-	0.15***
lnInvFixCap	0.26	0.51	0.22	0.34	-	–	0.47*	0.49*	0.52*	0.49*	0.49***	0.50*
shFCInvPHealth	0.03**	-	0.05**	0.04**	-	–	0.02**	-	0.04***	0.04***	-	0.04***
shPubGradOEI	-	−0.04**	−0.04*	−0.06**	−0.07**	−0.05**	−0.05**	−0.05**	−0.05**	−0.04***	−0.05***	−0.04*
lnPostdocStud	−0.23	-	-	-	-	–	−0.12	-	−0.10	-	-	-
lnInnOutSpill	0.25***	0.26***	0.28***	0.24***	0.31***	0.29***	0.30*	0.19***	0.28***	0.25***	0.25***	0.24***
lnGPatentSpill	−0.33***	−0.26***	−0.31***	−0.29***	−0.36***	−0.33***	−0.44*	−0.26*	−0.33***	−0.31***	−0.31*	−0.31*
lnGRPpc	−0.14	−0.32*	−0.17	-	-	-	-	-	-	-	-	-
lnGRP	-	-	-	-	-	-	0.27	0.39	0.14	-	-	-
R^2	0.89	0.88	0.89	0.88	0.87	0.88	0.88	0.88	0.89	0.88	0.87	0.88
R^2-adjusted	0.88	0.87	0.88	0.88	0.87	0.88	0.88	0.87	0.88	0.88	0.87	0.88
Sargan Test P-value	1.00	1.00	0.99	1.00	1.00	1.00	1.00	1.00	0.99	1.00	1.00	1.00

***,**,* Statistically significant, respectively, at the 1, 5 and 10% level

For the hypothesis of knowledge diffusion from neighbor regions the positive spillover effect of the innovative output is contrasted with the negative effect of the number of granted patents. The former finding can be considered as plausible: a good or a service is a data carrier itself, and being exported can transfer knowledge to companies in neighbor regions. On the contrary, the latter effect is a surprising result, as empirical evidence from the US and Western Europe confirms patents as an instrument of knowledge diffusion (Bacchiocchi and Montobbio 2009). These two effects are found to be strongly significant and are included with all IC in the final model.[15]

One might interpret this as a result of the property rights policy: technologies or goods patented in one region are protected from copying in neighbor regions. In this case, the technologies need to be either very close or even the same, which is a very strong assumption. Another explanation for this is a concentration of Russian innovative companies in 'special economic zones' (RusSEZ) with certain tax reliefs and further benefits for innovative companies (see Dezhina *et al.* (2008)). By this, regions within these zones absorb innovative production from their neighbors instead of transferring knowledge.

Based on this result one can conclude that knowledge is regionally bound in Russia and particular measures to develop inter-regional knowledge diffusion are worth to undertake. This technology transfer can be supported via knowledge diffusion centers transferring knowledge between research and industry and promoting cooperation between regions. For this reason in Russia already exists an association of public scientific centers. Following the example of the EU with the programmes like CORDIS and Innovative Relay Centres, this association can build a federal database of innovative companies with their business and technical solutions assisting other firms to find partners and benefit from this knowledge. Besides, due to a large variety of agencies, committees and ministries stimulating innovative activities in Russia (Dezhina *et al.* 2008), more coordination between these entities is required.

For the hypothesis on financial performance the only significant indicator selected is the share of aggregated net profits in GRP. However, to draw any concrete conclusions, the hypothesis must be tested on microdata.

Among indicators on infrastructure the density of rail roads, the turnover of goods by means of motor roads and the investments in fixed capital have partial positive effects on the dependent variable. Though, their inclusion in the final model is dependent on the assumption on the initial conditions process. The only exception for this are the investments in public health services with a

[15]Remember that these results are obtained with the neighborhood concept we have chosen. It is also relevant for further examination to test this hypothesis with a different concept, allowing, e.g., for knowledge spillovers between Moscow and St. Petersburg.

significant positive influence on the innovative output. So far, we do not have an exhaustive explanation for this. Thus, we find some evidence for infrastructure indicators to have an impact on the innovative performance, but until now, it is impossible to draw more specific conclusions.

Finally, there is no significant indicator identified for the hypotheses on impact of public, municipal and private forms of ownership. Neither any control variable is found to be significant.

As a result, our study supports implementation of the catching-up strategy (Polterovich 2009) imitating foreign state-of-the-art technologies and increasing own innovative potential. For this a series of measures increasing the efficiency of FDI, education system and knowledge diffusion is discussed.

3.6 Conclusions and Outlook

In this paper the innovative performance of Russian regions is analyzed. The innovative process is described in numerous studies with sometimes contradictive results. As there is no generally accepted model available encompassing all the factors of interest, a generalized log-linear model based on the regional data of Rosstat is suggested.

We optimize the model structure by selecting only those variables, which are relevant according to the information criteria. It is demonstrated that the corresponding optimization problem is complex due to the large discrete search space. Therefore, no classical optimization methods can be applied.

To deal with this problem two heuristic optimization methods (Threshold Accepting and Genetic Algorithms) able to find an optimum or at least a very good result in terms of the IC value are suggested. Comparing the heuristics we argue that for this particular problem for a given CPU time, GA provide marginally better results in terms of the mean, minimum values and variance.

One problem that becomes obvious for our application is the asymptotic behavior of IC. A sample size of 600 observations is too small for IC to identify the 'true' model. Relaxing the 'group constraint' (at least one regressor from each hypothesis to be included) reduces the number of false variables in the final model substantially (especially for BIC). However, the problem of relevant variables possibly not selected by IC remains open.

Taking the unobserved heterogeneity and possible endogeneity of regressors into account, we construct a dynamic model specification using GA and estimate it with the system GMM method. Based on this study a series of hypotheses on factors driving innovations is tested and possible ways to improve the impact of those factors are discussed.

In conclusion, we argue that the heuristic methods based on the IC are effective methods of model selection also for complex panel data models. In the future we will enhance our model selection procedure enabling to distinguish between strict exogenous, predetermined and endogenous variables simultaneously. For further study remains also the incorporation of the optimal choice of moments that could efficiently explore linear GMM moment restrictions and reduce the variance of the estimator. Application of our method on a different dataset, in particular with a larger sample size and with more accurate proxies on the market competition, investment climate and other hypotheses, is also of interest.

Acknowledgements Valuable comments and suggestions from Christian Borgelt and Frank Windmeijer are gratefully acknowledged. All remaining shortcomings are our responsibility.

3.7 Appendix

3.7.1 Description of the Data Used

Table 3.5: List of tested explanatory variables.

	Product market competition	
shInnovOrg	Share of organizations which produce innovative output	
lnRDCost	Log of internal R&D costs	
shInterCostLab	Share of expenditure on labor remuneration in internal R&D costs	
shInterCostSTax	Share of expenditure on social taxes in internal R&D costs	
shInterCostEqp	Share of expenditure on machines and equipment in internal R&D costs	
shInterCostMat	Share of expenditure on materials	
shOrgRDactiv	Share of organizations which conduct R&D activities	
lnApplPatent	Log of number of applications for patents	
lnGrantPatent	Log of number of granted patents	
lnAdvProdTech	Log of number of advanced production technologies used	
	Infrastructure	
lnRevCommServ	Log of revenues from communication services per capita	
lnNumPhonepc	Log of number of telephones per thousand inhabitants	
shMobPhoneSubs	Share of mobile telephone subscribers in regional population	
lnRailRoadDen	Log of kilometers of railroads of general use per 10000 km^2	
lnMotorRoadDen	Log of kilometers of motor roads of general use per 1000 km^2	
shCreditOrg	Share of credit organizations in the total number of organizations	
shCreditOrgAff	Share of credit organizations' affiliates in the total number of organizations	
lnTurnRailRoad	Log of turnover of goods by means of railroads	
lnTurnMotorRoad	Log of turnover of goods by means of motor roads	
lnInvFixCap	Log of fixed capital investments (FCInv's)	
shFCInvIndProd	Share of FCInv's in industrial production	distribution according to economic activity
shFCInvAgricul	Share of FCInv's in agriculture	
shFCInvConstr	Share of FCInv's in construction	
shFCInvTrans	Share of FCInv's in transport	
shFCInvComm	Share of FCInv's in communication	
shFCInvTrade	Share of FCInv's in trade	
shFCInvPHealth	Share of FCInv's in public health	
shFCInvEduc	Share of FCInv's in education	
	Scale of production	
shSME	Share of small companies	
shSMEoutput	Share of SME's output in gross regional product (GRP)	
	Form of ownership	
lnFDI	Log of foreign direct investments	
lnFPI	Log of foreign portfolio investments	
lnFIE	Log of foreign investments not elsewhere specified	
shPubInvFixCap	Share of public investments in FCInv's	distribution according to ownership structure
shMunInvFixCap	Share of municipal investments in FCInv's	
shPriInvFixCap	Share of private investments in FCInv's	
shPrivPubMunOrg	Share of privatized public and municipal organizations	
shEqtInvFixCap	Share of equity in FCInv's	distribution according to source of finance
shBCrInvFixCap	Share of bank credits in FCInv's	
	Economic performance	
shAggrNetPinGRP	Share of aggregated net profit of companies in GRP	
avNetProfit	Average net profit of companies in roubles (m)	

lnAggrCredDinRu	Log of aggregated credit debts of companies in roubles
lnAggrCredDinFX	Log of aggregated credit debts of companies in foreign currency
avCredDinRu	Average credit debts of companies in roubles (m)
avCredDinFX	Average credit debts of companies in foreign currency (in roubles, m)
	Knowledge spillovers I
shPubHighSGrad	Share of public high-school graduates in total population (TP)
shPrivHighSGrad	Share of private high-school graduates in TP
shPubGradOEI	Share of graduates from other public educational institutions in TP
shPrivGradOEI	Share of graduates from other private educational institutions in TP
shPubHStoAllHS	Share of public graduates in all high-school graduates
shPubtoAllGrad	Share of public graduates in all graduates from other educational institutions
lnRDstaff	Log of the number of employees in R&D departments
lnDocStudent	Log of number of doctoral students
lnPostdocStud	Log of number of postdoctoral students
shExpRWorld	Share of export to the rest of the world relative to GRP
shImpRWorld	Share of import to the rest of the world relative to GRP
shExpCIS	Share of export to the CIS countries relative to GRP
shImpCIS	Share of import to the CIS countries relative to GRP
lnExpNumofTech	Log of number of contracts for export of technologies
lnExpValofTech	Log of value of contracts for export of technologies
lnExpEarnofTech	Log of annual earnings of contracts for export of technologies
lnImpNumofTech	Log of number of contracts for import of technologies
lnImpValofTech	Log of value of contracts for import of technologies
lnImpEarnofTech	Log of annual earnings of contracts for import of technologies
	Knowledge spillovers II
lnInnOutSpill	Log of value of innovative products in neighbor regions (NR)
lnRDCostSpill	Log of internal costs on R&D in NR
shOrgRDactSpill	Share of organizations which conduct R&D activities in NR
lnGPatentSpill	Log of number of granted patents in NR
lnAdvPrTSpill	Log of number of advanced production technologies used in NR
shPHighSGSpill	Share of public high-school graduates in the total population in NR
lnRDstaffSpill	Log of the number of employees in R&D departments in NR
	Control variables
lnGRP	Log of GRP in current prices
lnGRPpc	Log of gross regional product per capita in current prices
lnRevConsBudg	Log of revenues of regional consolidated budgets
lnValLTAssets	Log of value of regional long term assets
shEmplPop	Share of employable population in total population
shUrbanPop	Share of urban population in total population
shUnEmplPop	Share of unemployed population (relative to employable population)
lnNumofApplLF	Log of number of applications for labor force
lnNumofRegCrime	Log of number of registered crimes per 100 000 inhabitants

Table 3.6: List of regions included in the dataset.

1.	Altai Krai	26.	Kurgan Oblast	51.	Sakha (Yakutia) Republic
2.	Amur Oblast	27.	Kursk Oblast	52.	Republic of North Ossetia-Alania
3.	Arkhangelsk Oblast	28.	Leningrad Oblast	53.	Republic of Tatarstan
4.	Astrakhan Oblast	29.	Lipetsk Oblast	54.	Rostov Oblast
5.	Belgorod Oblast	30.	Magadan Oblast	55.	Ryazan Oblast
6.	Bryansk Oblast	31.	Moscow Oblast	56.	Samara Oblast
7.	Vladimir Oblast	32.	Murmansk Oblast	57.	Saratov Oblast
8.	Volgograd Oblast	33.	Nizhny Novgorod Oblast	58.	Sakhalin Oblast
9.	Vologda Oblast	34.	Novgorod Oblast	59.	Sverdlovsk Oblast
10.	Voronezh Oblast	35.	Novosibirsk Oblast	60.	Smolensk Oblast
11.	Moscow	36.	Omsk Oblast	61.	Stavropol Krai
12.	Saint Petersburg	37.	Orenburg Oblast	62.	Tambov Oblast
13.	Jewish Autonomous Oblast	38.	Oryol Oblast	63.	Tver Oblast
14.	Ivanovo Oblast	39.	Penza Oblast	64.	Tomsk Oblast
15.	Irkutsk Oblast	40.	Perm Krai	65.	Tula Oblast
16.	Kabardino-Balkar Republic	41.	Primorsky Krai	66.	Tyumen Oblast
17.	Kaliningrad Oblast	42.	Pskov Oblast	67.	Udmurt Republic
18.	Kaluga Oblast	43.	Republic of Adygea	68.	Ulyanovsk Oblast
19.	Kamchatka Oblast	44.	Republic of Bashkortostan	69.	Khabarovsk Krai
20.	Karachay-Cherkess Republic	45.	Buryat Republic	70.	Khanty-Mansi Autonomous Okrug
21.	Kemerovo Oblast	46.	Republic of Dagestan	71.	Chelyabinsk Oblast
22.	Kirov Oblast	47.	Republic of Karelia	72.	Chita Oblast
23.	Kostroma Oblast	48.	Komi Republic	73.	Chuvash Republic
24.	Krasnodar Krai	49.	Mari El Republic	74.	Yamalo-Nenets Autonomous Okrug
25.	Krasnoyarsk Krai	50.	Republic of Mordovia	75.	Yaroslavl Oblast

3.7.2 System GMM Estimation Technique

Here we will state the assumptions of the system GMM estimators applied. First, μ_i and ν_{it} are independently identically distributed so that

$$E(\mu_i) = 0; \; E(\nu_{it}) = 0; \; E(\nu_{it}\mu_i) = 0; \; i = 1,...,N; \; t = 2,\ldots,T \tag{3.10}$$

and there is lack of serial correlation, but not necessarily independence over time:

$$E(\nu_{it}\nu_{is}) = 0; \; i = 1,...,N; \; t = 2,\ldots,T; \; \forall t \neq s. \tag{3.11}$$

Following Blundell and Bond (1998) we also make the standard assumption concerning the initial conditions y_{i1}:

$$E(y_{i1}\nu_{it}) = 0; \; i = 1,...,N; \; t = 2,\ldots,T. \tag{3.12}$$

Conditions (3.10), (3.11), (3.12) are sufficient for the following $(T-1)(T-2)/2$ linear moment conditions to be valid:

$$E[y_{i,t-2}\Delta(\mu_i + \nu_{it})] = 0; \; t = 3,\ldots,T, \tag{3.13}$$

where $\Delta(\mu_i + \nu_{it}) = \Delta u_{it} = u_{it} - u_{i,t-1} = \Delta y_{i,t} - \widehat{\delta}\Delta y_{i,t-1} - \Delta x'_{i,t}\widehat{\beta}$. Introducing lagged values of y_{it} as instruments, we estimate δ in first-differences for datasets with a time-series dimension $T \geq 3$. But as these estimations are biased (Blundell and Bond 1998), we need to make an additional mild stationarity assumption about the initial conditions y_{i1} allowing the use of an extended 'system GMM' estimator that uses lagged differences of y_{it} as instruments for equations in levels. This stationarity condition on y_{i1} requires $E[(y_{i1} - \frac{\mu_i}{1-\delta})\mu_i] = 0$ for $i = 1,\ldots,N$, so that y_{it} converges towards its mean $\frac{\mu_i}{1-\delta}$ for each region from period $t = 2$ onwards. This yields the condition:

$$E[\Delta y_{i,t-1}\mu_i] = 0; \; i = 1,\ldots,N. \tag{3.14}$$

If (3.10), (3.11), (3.12) and (3.14) hold, the additional $(T-1)(T-2)/2$ moment conditions are valid:

$$E[\Delta y_{i,t-1}(\mu_i + \nu_{it})] = 0; \; t = 3,\ldots,T. \tag{3.15}$$

Together the moment conditions on equations in first-differences (3.13) and on equations in levels (3.15) yield the system GMM estimator (Blundell and Bond 1998).

We do not assume that the explanatory variables x'_{it} are all strictly exogenous, but either endogenous or predetermined. x'_{it} might be endogenous in the sense that they are correlated with ν_{it} and earlier shocks, but not with subsequent ones:

$$E(x'_{it}\nu_{is}) \neq 0; \;\; i = 1, \ldots, N; \;\; \forall s \leq t. \tag{3.16}$$

x'_{it} are predetermined if in addition it is assumed that there is no correlation between x'_{it} and ν_{it}. Then, conditions (3.12), (3.13), (3.14) and (3.15) are also true for x'_{it}, allowing to include $x_{i,t-2}$ and $\Delta x_{i,t-1}$ as valid instruments.

Then we can obtain $\widehat{\delta}$ and $\widehat{\beta}$ in a two-step procedure. In the first step, we transform (3.4) in the first-difference equation:

$$y_{it} - y_{i,t-1} = \delta(y_{i,t-1} - y_{i,t-2}) + (x'_{it} - x'_{i,t-1})\beta + (\nu_{it} - \nu_{i,t-1}), \tag{3.17}$$

and perform the Generalized Least Squares (GLS) preliminary one-step consistent estimation:

$$\begin{pmatrix} \widehat{\delta}_1 \\ \widehat{\beta}_1 \end{pmatrix} = \left[\begin{pmatrix} \Delta y_{t-1} \Delta X_t \\ y_{t-1} \quad X_t \end{pmatrix}' W (\Sigma_{i=1}^{N} W_i' G W_i)^{-1} W' \begin{pmatrix} \Delta y_{t-1} \Delta X_t \\ y_{t-1} \quad X_t \end{pmatrix} \right]^{-1} \cdot$$
$$\cdot \left[\begin{pmatrix} \Delta y_{t-1} \Delta X_t \\ y_{t-1} \quad X_t \end{pmatrix}' W (\Sigma_{i=1}^{N} W_i' G W_i)^{-1} W' \begin{pmatrix} \Delta y_t \\ y_t \end{pmatrix} \right], \tag{3.18}$$

where $W = \left[W_1', \ldots, W_i' \right]'$ is a matrix of instruments for all regions W_i and ΔX is the stacked $N(T-2) \times k$ matrix of observations on $\Delta x'_{it}$:

$$W_i = \begin{pmatrix} W_i^d & 0 \\ 0 & W_i^l \end{pmatrix}, \tag{3.19}$$

where W_i^d contains instruments in first-differences, and W_i^l contains non-redundant instruments in levels. For endogenous x'_{it}, W_i has the form:[16]

$$W_i^d = \begin{pmatrix} [y_{i1}, x'_{i1}] & 0 & \cdots & 0 \\ 0 & [y_{i1}, y_{i2}, x'_{i1}, x'_{i2}] & \cdots & \vdots \\ \vdots & \ddots & \ddots & \vdots \\ 0 & \cdots & \cdots & [y_{i1}, \ldots, y_{i,T-2}, x'_{i1}, \ldots, x'_{i,T-2}] \end{pmatrix},$$

[16]For predetermined x'_{it} one more lag in W_i^d and first-differences from $\Delta x'_{i3}$ up to $\Delta x'_{i,T}$ are included.

$$W_i^l = \begin{pmatrix} [\Delta y_{i2}, \Delta x_{i2}'] & 0 & \cdots & 0 \\ 0 & [\Delta y_{i3}, \Delta x_{i3}'] & \cdots & \vdots \\ \vdots & \ddots & \ddots & \vdots \\ 0 & \cdots & \cdots & [\Delta y_{i,T-1}, \Delta x_{i,T-1}'] \end{pmatrix}.$$

For a more detailed analysis of the redundant moment conditions in dynamic panel data models see, e.g., Okui (2009). W_i is a $2(T-2) \times (k+1)(T-2)(T+1)/2$ matrix with k independent variables of x_{it}'. G is a $2(T-2)$ square matrix consisting of two $(T-2)$ square matrices and zero otherwise:

$$G = \begin{pmatrix} G^d & 0 \\ 0 & G^l \end{pmatrix}. \tag{3.20}$$

G^d is a square matrix of dimension $T-2$ that has twos in the main diagonal, minus ones in the first subdiagonals and zeros otherwise. G^l is the identity matrix of dimension $T-2$.

In (3.18) a stacked system $\binom{\Delta y_{t-1}}{y_t}$ is used, comprising $(T-2)$ equations in first differences and the $(T-2)$ equations in levels corresponding to periods $3, .., T$, for which instruments are observed (Blundell and Bond 1998).

After that we obtain $\Delta\widehat{\nu}_{i,t} = \binom{\Delta y_{i,t}}{y_{i,t}} - \widehat{\delta}\binom{\Delta y_{i,t-1}}{y_{i,t-1}} - \binom{\Delta X_{i,t}}{X_{i,t}}\widehat{\beta}$ as a $2(T-2) \times 1$ vector and construct the optimal weighting matrix of $W_i' \Delta\widehat{\nu}_{i,t}$:

$$\widehat{V_N} = \left[\Sigma_{i=1}^{N} W_i' (\Delta\widehat{\nu}_i)(\Delta\widehat{\nu}_i)' W_i\right]^{-1}, \tag{3.21}$$

where the generalized Moore-Penrose inverse (MPI) is used (Penrose 1956). The MPI is applied because the matrix in (3.21) might be close to singular, triggering inaccurate results of the classical matrix inversion.

Inserting (3.21) in (3.18), we estimate the final coefficients. When $\nu_{i,t}$ are *i.i.d.*, the one-step and two-step estimators are asymptotically equivalent.

The system GMM estimator that uses instruments both in lags and in levels would loose its consistency if in fact the assumptions made in (3.13) and (3.15) were not fulfilled. To test the validity of the instrumental variables we perform the Sargan test (ST):

$$S_t = \Delta\widehat{\nu}' W \left[\Sigma_{i=1}^{N} W_i' (\Delta\widehat{\nu}_i)(\Delta\widehat{\nu}_i)' W_i\right]^{-1} W' \Delta\widehat{\nu} \sim \chi^2_{p-h}, \tag{3.22}$$

where $\widehat{\nu}'$ denote residuals from the two-step estimator, p is the number of columns in W_i (number of moment conditions) and h is the number of explanatory variables. The null-hypothesis of the ST asserts that the orthogonality conditions mentioned in (3.13) and (3.15) hold.

Chapter 4

Heuristic Model Selection for Leading Indicators in Russia and Germany

Ivan Savin, Peter Winker

The final publication is available at http://www.oecd.org/std/jbcma

4.1 Introduction

For many OECD countries, indicators based on business surveys became a standard tool for the analysis and forecasting of business cycles and for assessing models of expectation formation. Recent contributions and reviews on this literature include Chamberlin (2007), Claveria *et al.* (2007) and Goggin (2008). These indicators are also important as information on the current status of the economy given the publication delay of data on GDP or – with a smaller time lag – industrial production (Hüfner and Schröder 2002, p. 317). Such information is of particular interest in periods of a rapid changing economic environment like the financial market crisis starting in 2008 or for economies in transition. For these reasons, business tendency surveys (BTSs) are conducted in an increasing number of countries mostly on a monthly basis. While in the late 1960s there were about 30 indicators from such surveys, which were available only for the main OECD countries, today, the number of these indicators exceeds 300, which cover more than fifty countries (Nardo 2003, p. 645) including also economies in transition like Brazil, China, India and Russia (Nilsson and Brunet 2006).

For Germany – as for other industrial countries – there is a large body of empirical evidence that models forecasting industrial production (IP) or real

GDP, which include BTSs, outperform forecasts of univariate time series models (Hüfner and Schröder 2002, Kholodilin and Siliverstovs 2006). Similar evidence can be found for other OECD countries (Ozyldrim *et al.* 2010, Carstensen *et al.* 2011). For Russia, in contrast, there is less evidence on the usefulness of BTSs in forecasting IP or other aggregate measures of output (one exception is presented by Nilsson and Brunet (2006)). Possible reasons for that may be a relatively short history of the surveys in the country and the fact that the expectations of Russian managers are 'poorly calibrated' in a sense that Russian managers systematically overestimate their business situation, resulting in overoptimistic forecasts (Aukutsionek and Belianin 2001).

We use recent BTS–data for Germany and Russia to compare the relative information content. For Germany we consider the business expectation indicator taken from the ifo business cycle survey. For Russia, a similar indicator of expected future production is provided by both the Institute for the Economy in Transition (IET) and the Federal State Statistics Service (Rosstat). Our focus is on comparing the contribution of these indicators in modelling dynamic adjustment and forecasting IP. The modelling framework used are VAR models for constructing impulse response functions and unconditional forecasts, while a naive forecast and AR(p) processes serve as benchmarks.

Using the VAR modelling framework, we can avoid certain problems like simultaneity bias. As the potential number of lags included based on information criteria (IC) is large, we compare full-specified VAR models with subset models obtained using a Genetic Algorithm (GA) (Alberto *et al.* 2010, Savin and Winker 2012) allowing for 'holes' in multivariate lag structures.

The focus of this work is threefold. First, it presents an application of a VAR modelling framework with BTSs on the example of Russia as an economy in transition and its comparison with Germany. Since none of the aforementioned authors analyzed BTS–indicators on their forecasting performance during a recession period,[1] the second contribution is to determine whether BTSs in both countries provide useful information before and during the recent crisis. Third, we illustrate the more flexible subset selection strategy based on GA (and the resulting advantages) in comparison to the standard approach for short time series data.[2]

The rest of this paper is structured as follows. Section 4.2 presents the database for both countries and the VAR modelling framework used. Results of model specification and forecasting with the standard (*'take all up to the p-th*

[1]The only exception for Germany is presented by Drechsel and Scheufele (2010).

[2]The latter focus is seen to be important for both countries since there are only few studies (Winker 1995, Winker 2000, Kapetanios 2007) demonstrating the application of the lag selection approach with 'holes' on VAR models and neither of them used Russian or German BTS data before.

lag') approach are reported in Section 4.3. Section 4.4 presents the heuristic algorithm for subset model selection and compares its results with the standard procedure. Section 4.5 concludes with a short summary and an outlook to further research.

4.2 Data and Modelling Framework

4.2.1 Data Description

The statistical indicators used in the following are obtained through BTSs conducted in Germany and Russia. The surveys are collected on a monthly basis and are usually available before the end of the respective month.[3]

The business cycle indicators for Germany are derived from the ifo business cycle test conducted first in an experimental phase in 1948. Since only the early 1960s, also the diffusion index is calculated (Strigel 1965). It is published on a regular base by the ifo Institute (Munich) since the early 1970s (Strigel 1997). The sample includes about 7000 enterprizes with half of them representing the manufacturing sector. The response rate is around 85%. The ifo BTSs, e.g., business climate and business expectation, have been extensively studied in the literature demonstrating their ability to improve forecasts of the German IP, real GDP and some monetary indicators (e.g., Langenmantel (1999), Flaig (2003), Mittnik and Zadrozny (2005)).

For Russia we consider the information collected both by the Institute for the Economy in Transition (IET) and the Federal State Statistics Service (Rosstat), combining them in one data set. This is necessary due to the limited data availability with regard to the BTSs. From IET the data is available for the period 02/1999–09/2009,[4] whereas the data of Rosstat is available since 07/1997, but has a structural break for the entire year 2005.[5] Hence, for the period 02/1999-12/2005 we use the IET data and for 01/2006–09/2009 Rosstat data. The coverage of the IET survey is relatively wide: with 227 administrative regions and 307 towns. Data are sampled for 1200 enterprizes from all major industries with a response rate of about 65% (Tsukhlo 2005). The coverage of the Rosstat is even larger (over 4000 companies) from all major industries and all Russian regions.[6]

Because of the limited data availability for Russia we consider the period 02/1999–09/2009, leaving one or two years for an out–of–sample forecast anal-

[3] In Russia the data is published a week after the end of the month.

[4] Officially, these BTSs are collected since 03/1992 to date.

[5] Due to a change in classification systems no observations are available.

[6] See the Federal State Statistics Service's web site (http://www.gks.ru/).

ysis between 11/2006 and 09/2009. Thus, we consider the forecasting performance of our models both prior and during the crisis.[7] By restricting our data set in this way, we avoid to take into account the potential structural breaks due to reunification in 1990 and introduction of the Euro in 1999 for Germany and the period of financial crisis in Russia in 1998.

Considering Russia and Germany allows us to compare the relative performance of BTSs for a country in the process of transition, which collects this type of data only for around ten years, with an industrial country being one of the pioneers in using business survey data.

The business expectation from the ifo survey refers to a planning horizon of six months according to the questionnaire. The indicator of future production for Russia (both IET and Rosstat) is calculated in a similar way, but with a time horizon of only three months. Obviously, this difference in the time horizon considered might result in differences of the dynamic adjustment in a VAR model setting (see Section 4.3).

We also refer to the business climate indicator in our analysis. It is calculated as an arithmetic mean of the results (balances) for the three questions related to production, total demand and finished goods stocks tendencies (with negative weight) for Russia, while a geometric mean of the present and future business situation is used for Germany (for further details see OECD (2003)). Thus, in contrast to the business expectations, these indicators can be referred to as composite leading indicators (see among others, Goldrian (2007)). In the upper plot of Figure 4.1 the monthly IP growth rates (left axis) together with the business expectations and business climate perception (right axis) for Germany (left panel) and Russia (right panel) are shown. It can be observed that BTSs in both countries were predicting the drop in IP for 2008. Considering yearly growth rates (lower plot of the same figure) one can conclude that for Germany leading indicators started to predict a drop in IP even earlier. Furthermore, it is remarkable that the business expectations indicator for Russia is almost always higher than the business climate index (upper left plot), whereas for Germany there is no systematic margin between the two indicators.

An important difference between the Russian and German data is found with regard to the treatment of seasonal variation. While ifo asks its correspondents to provide seasonally adjusted estimates, both IET and Rosstat do not require this correction. As a result, the BTS indicators in Russia have a strong seasonal component (see Figure 4.1, right panel, dashed lines). Furthermore, the figure exhibits that the seasonal pattern changes over time and becomes more pronounced from 2004 onward. Interestingly, if we consider data collected by

[7]Note that starting from the fourth quarter of 2008 GDP and IP growth rates for Russia and Germany become negative.

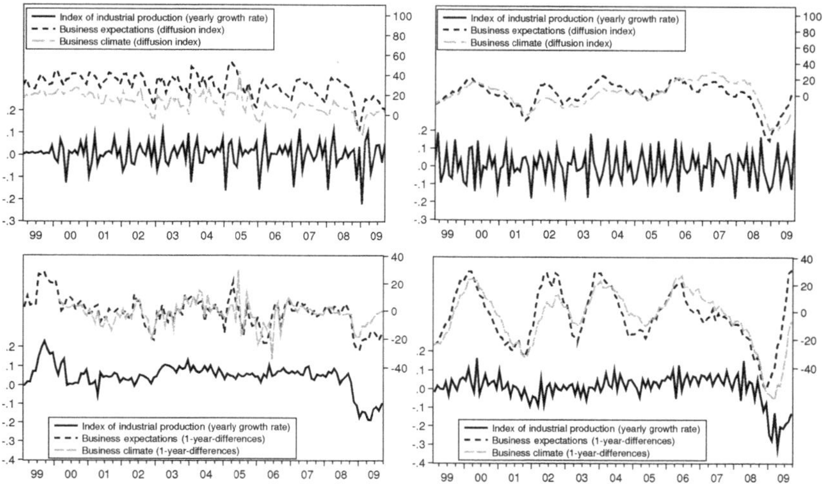

Figure 4.1: BTSs and IP growth rates for Germany and Russia

the OECD available only up to 06/2007 (meanwhile the OECD statistical service terminated the data collection for Russia), a similar seasonal pattern is also observed for the period from 1995 to 2004. So far, we are not able to explain this surprising change in the seasonal pattern.[8]

Because the data on the Russian BTSs is not seasonally adjusted, we will also consider the IP indices for both countries as not seasonally adjusted. First, it will provide a more consistent basis for the comparison between the two countries. Second, instead of using some a priori methods of seasonal adjustment, which might have unwelcome side effects, e.g., the problem of spurious regression (for more details, see among others Meyer and Winker (2005)), introducing seasonal dummies in the model allows also to analyze this component and its effect on the dynamic adjustment process.[9]

In addition, due to the necessity to merge two data sets for Russia, we account for a potential structural break in our VAR models for Russian data and for this reason add a shift dummy consisting of zeros until 12/2005 and ones

[8]Our exchange with IET regarding this finding did not result in a satisfying explanation. Nevertheless, it will have to be taken into account for the empirical model.

[9]Alternatively, we also consider year-on-year growth rates both for IP and BTSs without including seasonal dummies and trying in this way to relax the degrees of freedom problem. In general, however, this does not change any major conclusion (contribution of BTSs before and during the crisis, comparison between standard and GA model selection strategies) stated below. Results are available upon request.

starting from 01/2006 onwards. Furthermore, we include its first difference and lagged values of the first difference (the same number of lags as for the corresponding VAR model) to capture the dynamics given in the models. Also, interaction terms between the seasonal dummies and the shift dummy are included in the models for Russia.[10] This results in a more flexible model accounting for the change in the seasonal structure for Russian BTSs.

As regards IP indices, for Germany, we use the data provided by the Federal Statistical Office with base year 2005. The index accounts for just under 80% of total production in the manufacturing sector and covers enterprizes also in former East Germany.[11] For Russia we consider the data provided by Rosstat with base year 2002 and also a broad coverage of firms.

Using monthly IP indices (instead of quarterly GDP) has certain advantages. First, available at monthly frequency IP requires no aggregation of the BTSs and, hence, no loss of information. Second, the number of observations for IP is higher increasing the available degrees of freedom.

However, it should be noted that for Germany IP covers less than 50% of total economic activity (Kholodilin and Siliverstovs 2006, p 240). For Russia this figure is even closer to 30% (Domsch and Lidokhover 2007). Moreover, in both countries the service sector becomes increasingly relevant, which might exhibit a different dynamic in comparison to IP. In contrast, as it was mentioned, the BTS indices cover all main industries of the two economies, including IP, services and trade (Tsukhlo 2005). Thus, we acknowledge some differences in the structure of the IP index we forecast and the BTSs we use.

4.2.2 Model Specification

Given that no a priori information is available on the mutual dynamic adjustment of IP and leading indicators, a vector autoregressive (VAR) model is chosen as a reduced form modelling framework:

$$\begin{aligned} Y_t &= \sum_{i=1}^{p} \beta_{1,i} Y_{t-i} + \sum_{j=1}^{q} \gamma_{1,j} E_{t-j} + \sum_{d=1}^{12} \delta_{1,d} S_{d,t} + \varepsilon_{1,t} \\ E_t &= \sum_{i=1}^{p} \beta_{2,i} Y_{t-i} + \sum_{j=1}^{q} \gamma_{2,j} E_{t-j} + \sum_{d=1}^{12} \delta_{2,d} S_{d,t} + \varepsilon_{2,t}, \end{aligned} \tag{4.1}$$

[10] Significance of the interaction terms for the corresponding VAR models is confirmed via the F-test at the 1% significance level.

[11] See the Federal Statistical Office's web site (http://www.destatis.de).

where Y stands for the monthly IP growth rate, E for the BTS indicator and S_d for seasonal dummies. Our focus is on identifying the dynamic dependence between the variables of interest and whether information from BTSs can improve forecast accuracy for IP. When evaluating the forecasting performance, we will consider a naive forecast (NF) as one benchmark. The NF forecasts are modeled as mean average values of the IP growth rates over the respective estimation period, i.e., the h–step–ahead–forecast $\widehat{Y}_{t+h}$ made at time t is the average of the variable Y in the period between 1 and t:

$$\widehat{Y}_{t+h} = \frac{1}{t} \sum_{\eta=1}^{t} Y_{\eta}. \tag{4.2}$$

This benchmark might not be hard to beat, but allows to compare the relative performance of alternative models. This is due to a similar performance of NF before and during the recession, which is not the case for, e.g., a simple random walk forecast. As alternative benchmarks, we consider AR(p)-processes to identify the contribution of the BTSs in the VAR models.

In a first step of the model specification, one has to consider possible non-stationarity of the time series. Thereby, it has to be taken into account that leading indicators refer to future changes in output. Therefore, the variables of interest are the growth rate of output, approximated by the monthly IP growth rate, and the BTSs. In Appendix 4.6.1, we summarize results of unit root tests. Although the results are not perfectly conclusive, they allow the assumption that both variables behave as stationary ones.

The lag length selection for the VAR models is based on the Schwarz (SIC) and the Akaike information criteria (AIC). The latter IC is asymptotically inconsistent (Shibata 1976), but penalizes free parameters less strongly than SIC, which is considered as an advantage for small sample sizes.

4.2.3 Goodness of Forecasts

To assess the quality of forecasts we consider two measures. First, the root mean squared forecast error (*RMSFE*)

$$RMSFE = \sqrt{\frac{1}{T_2 - T_1 + 1} \sum_{t=T_1}^{T_2} e_t^2}, \tag{4.3}$$

where e_t is the difference between the actual IP growth rate (y_t) and its forecast ($\widehat{y}_t$); T_1 and T_2 indicate the first and the last period in the out–of–sample forecast, respectively. The smaller the *RMSFE*, the better.

The second accuracy measure is 'Theil's U' (TU) (Theil 1966):

$$TU = \sqrt{\sum_{t=T_1}^{T_2} e_t^2} \Big/ \sqrt{\sum_{t=T_1}^{T_2} u_t^2}. \tag{4.4}$$

This measure can be interpreted as the ratio of the *RMSFE* of the analyzed model as compared to that of a reference model with forecast errors u_t.

We build a few series of out–of–sample forecasts (for one- and two–year periods) for the IP growth rates for both countries using VAR models. To investigate dynamics of the prediction errors, the so called 'rolling windows' method is used (for more information see Swansan (1998)). Doing this we either fix the estimation period, so that its length is the same for different overlapping samples - fixed estimation windows (FEW) - or let it increase always starting with 02/1999 - increasing estimation windows (IEW). We specify and estimate models for the forecasting exercise based on evolving estimation samples, while the analysis of stationarity, causality and impulse response functions is based on the full sample. Due to the limited number of observations after the structural break in Russia (01/2006), we include interaction terms only for a part of the simulations (forecasting samples) performing other forecasts without these flexible seasonal components.[12]

4.3 Results of the Standard VAR Model Selection

According to the standard approach to the model selection problem (see, e.g., Lütkepohl (2007)), one compares different lag lengths for the VAR model choosing the one with the best balance between model fit ($\widehat{\Sigma}$) and the total number of parameters to be estimated ($k = m^2p + md$).[13] Thereby, in this section we impose the standard restriction that all lags up to a predefined order will be included (*'take all up to the p-th lag'* approach) and that this lag length is the same for all variables in all equations ($p = q$ in the notation of equation (4.1)). Due to the monthly data frequency, the maximum lag length (p_{max}) is set to 13. The results are provided in Table 4.8 in Appendix 4.6.2.

In the following, the resulting VAR models are compared with regard to their out–of–sample performance. In particular, we use a 1–step–ahead forecast that employs only actual observations. First, we consider results with increas-

[12] Due to the same reason we modify the size of the fixed estimation samples for simulations with and without the interaction terms (101 and 93 observations, respectively).

[13] $m = 2$ is the number of endogenous variables, p is the number of lags and $d = 12$ and 27 is the number of exogenous variables for Germany and Russia, respectively.

ing estimation windows (IEW) for an evaluation period of one and two years (Table 4.1), respectively. The German IP is forecasted best by the VAR model with business expectations specified according to SIC, while the VAR with the business climate based on AIC is the second-best choice. These models have an *RMSFE* smaller than the benchmarks. Similarly, the VAR forecasts for Russia specified by SIC are more accurate than the ones based on AIC. The improvement is of a similar scale as in Germany.[14]

Table 4.1: Forecasting performance (RMSFE) of the VAR models

Model specification		Germany	Russia	Germany	Russia
		10/2007–09/2009		10/2008–09/2009	
NF forecast		10.0544	7.7724	10.2511	8.9200
AR forecast	AIC	6.3691	4.6519	8.0974	6.1679
	SIC	6.3898	4.7360	8.2208	6.3296
Business expectations	AIC	5.5188	5.3519	6.9972	6.2860
	SIC	4.9840	3.8418	6.0813	5.0850
Business climate	AIC	5.3051	4.0459	6.6216	5.4377
	SIC	5.4239	3.8520	6.7384	5.1057

Next, to analyze goodness of the resulting forecasts for other periods, we consider *TU* relative to the NF forecast for overlapping windows within 11/2006–09/2009. Results for SIC[15] with IEW are given in Figure 4.2 for one-year periods (for two–year periods and FEW see Appendix 4.6.4). On the x-axis the forecasted period is given, while the y-axis indicates values of Theil's U.[16] While prior to the end of 2008 (when the crisis has become pronounced also in IP indices) the values of Theil's U vary approximately between 0.3–0.5, the relative performance of the VAR models decreases dramatically afterwards.

The question, however, is whether the BTS indicators are useful in producing those forecasts (both prior and during the recession). To this end, the VAR models need to be compared with a different benchmark. The AR(p)-processes are most suitable for this comparison, because the contribution of the BTSs as additional indicators is expected to become evident. Nonetheless, as one can

[14]We also tested the Mincer–Zarnowitz regression (Mincer and Zarnowitz 1969) to ensure that the produced forecasts are not biased. For this purpose we consider t-tests for separate coefficients and p-values of the Wald test as a joint test. Results for the period 10/2008–09/2009 with IEW are presented in Table 4.9 in Appendix 4.6.3 and provide some evidence that the respective hypotheses can not be rejected. Similar results are also obtained for other out–of–sample experiments.

[15]Results for AIC are available upon request.

[16]All figures containing results on rolling windows (Figures 4.2-4.5) present TUs in relation either to the NF or AR benchmark. See respective discussion for further explanation.

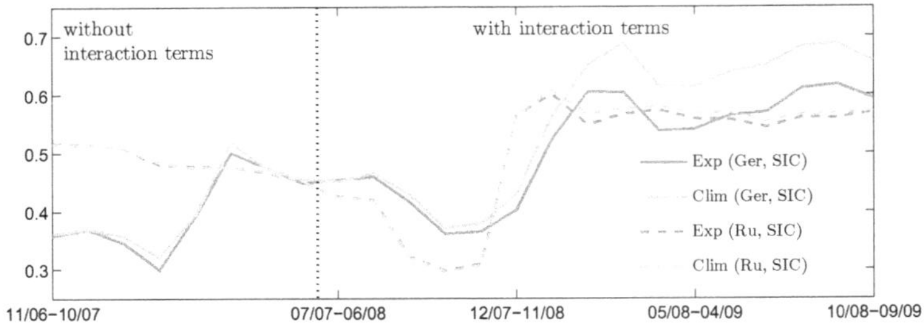

Figure 4.2: One–year forecast accuracy compared to NF with IEW (TU)

We denote in our VAR models Germany with 'Ger' and Russia with 'Ru', while 'Exp' and 'Clim' stand for business expectations and business climate.

see from Figure 4.3, the information content of BTSs is not clear, in particular for Russia. While for Germany the VAR models outperform the AR ones (with some minor exceptions), for Russia a high variance (that is even higher for AIC[17]) in the resulting performance is obtained. This is due to the variance in the lag length selected by IC resulting in a higher forecast error. For example, the difference in the forecast accuracy (in relation to AR) for the period 12/2007–11/2008 is about 10% lower than for the same length period shifted one month before or two months later.

Due to the high variance in the forecasting performance identified, the contribution of the BTSs is ambiguous. Nor their role prior and during the crisis is clear. Before one applies any significance tests on differences in out–of–sample performance between the VAR and AR models, we want to reduce the variance via an alternative model selection strategy allowing for 'holes' in the multivariate lag structures.

4.4 Heuristic Algorithm and Resulting Models

In the standard bivariate VAR model with 13 lags and seasonal dummies in our study, the rich lag structure requires k=76 parameters (52 lags and 24 dummies) to be estimated. For Russia this number is even higher (k=106) due to the shift dummies and interaction terms. Having only 128 observations (including a period for an out–of–sample forecast), this leaves only a very small number of degrees of freedom. If only a small portion of these parameters has non–zero

[17]These results are also available upon request, and for two–year periods and FEW, they can be found in Appendix 4.6.5. For the two–year forecasts it is also evident that the interaction terms improve the forecasting accuracy of Russian VAR models.

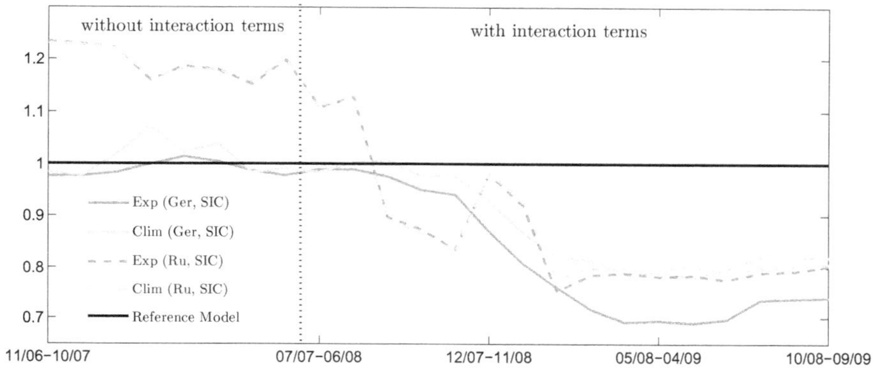

Figure 4.3: One–year forecast accuracy compared to AR with IEW (TU)

We denote in our VAR models Germany with 'Ger' and Russia with 'Ru', while 'Exp' and 'Clim' stand for business expectations and business climate.

coefficients in the real DGP, this results in a high estimation variance and forecast errors motivating us to use an alternative subset selection strategy allowing for 'holes' in multivariate lag structures.

Due to the discrete search space and 2^{m^2p} potential submodels we can neither apply standard optimization methods nor enumerate all possible solutions for high or even medium values of m and p (as, e.g., 2 and 13) comparing them based on IC. This cannot be done even using efficient branch–and–bound algorithms as described in Gatu *et al.* (2008).

In the last decade a large number of studies has been devoted to subset VAR selection methods. Among them are a sequential bottom–up (top–down) inclusion (deletion) of individual regressors (Lütkepohl 2007). Another option is to set certain model prior probabilities shrinking the parameter search space. The resulting model averaging strategies are well discussed in Kapetanios *et al.* (2008). However, these methods investigate only a certain portion of all potential submodels. An alternative is presented by the least absolute shrinkage and selection operator (Lasso) that was initially suggested by Tibshirani (1996) and later also applied to VAR models (Hsu *et al.* 2007). Due to the shrinkage parameter the method might exhibit a substantial estimation bias (Hsu *et al.* 2007, Savin forthcoming).

To tackle the complex integer optimization problem and, at the same time, to overcome a priori restrictions on the search space that might result in local optima, one can take advantage of optimization heuristics that mimic natural evolution processes (for an overview of the methods as well as for guidelines of their implementation see Gilli and Winker (2009)).

In our study we apply Genetic Algorithms (GA), which are a population–based heuristic that operate on a set of solutions (population).[18] The members in the GA population (chromosomes) are represented as bit strings, in which each position (gene) has two possible values (1 and 0), corresponding to included and not included regressors, respectively. In each generation GA replace parts of a population with new chromosomes (children) aimed to represent better solutions for a given problem.

Realizing the limitations of IC in our study (in particular, strong overfitting feature[19]), in the following we consider *RMSFE* as an objective criterion for GA. Thus, selecting a subset of lags, one estimates its out–of–sample performance on a 'pre-testing' period (which is twice as long as the actual out-of–sample period and precedes it) and uses the respective forecast error as a measure of goodness for a candidate solution. The intuition for this is that one can assume a model to demonstrate good out–of–sample performance, if the model well predicts the foregoing period. To this end, we concentrate on the selection within lags, but not within exogenous variables: First, one has not enough observations to estimate *RMSFE* on the 'pre-testing' period for interaction terms. Second, one avoids the risk to remove the controls due to their lower information content in comparison to lags.

The GA algorithm used in this study is very similar to the one described in Savin and Winker (2012). Some difference is that due to the smaller discrete search space we use a population of $p = 100$ solutions and preserve only one elitist chromosome (K^*). Besides, the number of generations $G = 100$ is sufficient for one restart (10 restarts for each setting).[20]

Comparing the full-specified models with subset models obtained using GA we consider both their in–sample and out–of–sample fit. While the in–sample analysis is made for the whole sample 02/1999–09/2009, the out-of-sample exercise is conducted (as described before) via rolling windows either of one or two years covering IP before and during the recession. In addition, dynamic forecasts based on the resulting GA subsets are constructed.

[18]In Savin and Winker (2012) this method is compared with a different heuristic (Threshold Accepting) and is found as more efficient for discrete subset selection problems.

[19]See (Savin and Winker 2011) for an extensive discussion.

[20]The corresponding CPU time for 100 generations using Matlab 7.11 on a Pentium IV 3.3 GHz requires approximately 1600s.

4.4.1 In–sample Fit

Two alternative model selection strategies, *'take all up to p-th lag'* and GA, are compared by means of the number of free parameters to be estimated (k)[21] and adjusted R^2 (for the IP growth rate's equation) (Table 4.2).

Table 4.2: In–sample fit of different models

Model specification			'all up to p-th lag'		GA	
			k	R^2_{adj}	k	R^2_{adj}
Germany	Business expectations	AIC	44	0.8692	34	0.8397
		SIC	32	0.8661		
	Business climate	AIC	40	0.8694	32	0.8462
		SIC	32	0.8539		
Russia	Business expectations	AIC	62	0.7981	32	0.7751
		SIC	54	0.7868		
	Business climate	AIC	62	0.7947	34	0.7727
		SIC	54	0.7860		

See Section 4.4.1 for explanation.

As one can see, for Germany GA include an intermediate number of parameters between AIC and SIC, whereas for Russia heuristics come up to a smaller subset than any IC demonstrating in both cases only a slightly lower in–sample fit than the standard lag selection strategy. Thus, omitting certain lags does not produce any substantial loss in the in–sample fit.

4.4.2 Static Forecasts and Significance Testing

The resulting out–of–sample fit is our main focus due to two reasons. First, a good out–of–sample performance gives a vital evidence in support of a particular model specification. Thus, a well specified model captures the given data set, but does not overfit it with too many predictors. Otherwise, forecasts based on the parameter estimates can be poor due to the estimates' variance. Second, for those interested in forecasting real production output it is more important that a specified model can produce good forecasts than fit to a certain data set.

To compare the forecasting performance of our VAR models specified based on the two alternative strategies, consider the resulting TUs (with NF as a benchmark) for one– and two–year out–of–sample periods (Table 4.3).

One can see that the GA forecasts based on *RMSFE* both for one- and two–year periods always outperform the forecasts based on the standard model selec-

[21] This includes both endogenous and exogenous variables.

Table 4.3: Forecasting performance of the VAR models

Model specification			Standard	GA	Standard	GA
			10/2007–09/2009		10/2008–09/2009	
Germany	Business expectations	AIC	0.5489	0.5075	0.6826	0.5964
		SIC	0.4957		0.5932	
	Business climate	AIC	0.5276	0.5027	0.6459	0.6282
		SIC	0.5395		0.6573	
Russia	Business expectations	AIC	0.6886	0.4834	0.7047	0.5506
		SIC	0.4943		0.5701	
	Business climate	AIC	0.5205	0.4802	0.6096	0.5363
		SIC	0.4956		0.5724	

See Section 4.4.2 for explanation.

tion approach via AIC. Moreover, they either outperform the standard strategy based on SIC (e.g., forecasts for Russia and business climate for Germany) or do not substantially differ in their performance (e.g., forecasts based on business expectations for Germany). This also holds true for other overlapping windows within the period 11/2006–09/2009 (see Figure 4.4).[22] Important to note is that, in contrast to the standard model selection, GA allow to reduce variation in the forecast accuracy. Thus, one avoids certain spikes in the *TU* dynamics as, e.g., for the period 04/2007–03/2008 for Germany or 01/2008–12/2008 for Russia.

Similarly, one can consider the resulting forecasting accuracy of our models in relation to the AR–benchmark (Figure 4.5).[23] Obviously, here we also obtain less variance in forecasting performance of our models. An important difference to the results of the standard strategy is that there is a much smaller difference between the AR and VAR forecasts for Russia.

Since the obtained *TU*s alone do not provide any evidence on statistical significance of the differences in forecasting performance between the AR and VAR models, we apply the Giacomini and White (2006) test (*GW* test henceforth) of conditional predictive ability.[24] The *GW* test is most suitable for our study for several reasons. First, it takes into account the effect of estimation uncertainty (West 1996) on the relative forecasting performance. Second, it can evaluate forecasts both for nested and nonnested models, which is particularly relevant for our case with an AR model being potentially nested within a VAR

[22]Results of the standard model selection strategy for business expectations via SIC (the best alternative identified in Section 4.3) are plotted for comparative reasons.

[23]In this case both AR and VAR model structures are optimized via GA.

[24]So far, available literature for Germany (Kholodilin and Siliverstovs 2006, Drechsel and Scheufele 2010) indicates that the difference between AR and VAR models in forecasting performance is significant only to some extent.

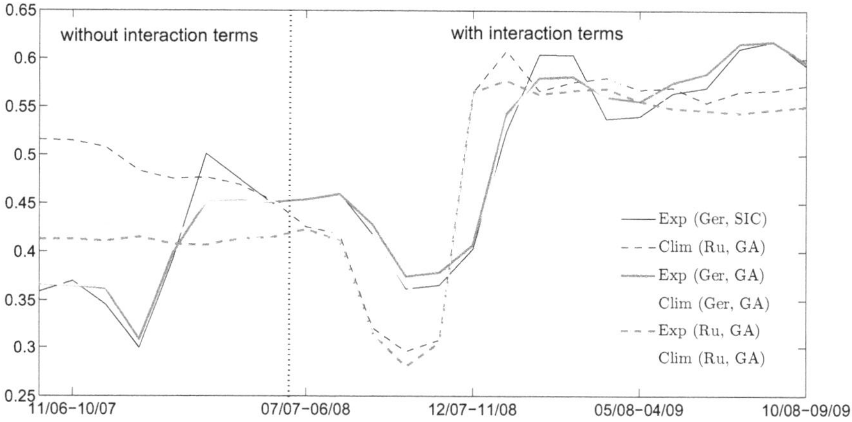

Figure 4.4: One–year forecast accuracy compared to NF with IEW (TU) via GA

We denote in our VAR models Germany with 'Ger' and Russia with 'Ru', while 'Exp' and 'Clim' stand for business expectations and business climate.

model for certain rolling windows. Third, presenting a generalized approach to existing studies in this field (as, e.g., Diebold and Mariano (1995)), the test allows for structural shifts at unknown dates (note the shift in the BTSs for Russia) and for both FEW and IEW approaches.

Defining $\Delta L_{m,t+1}$ as the difference of the loss function for the forecast based on the VAR model (e_{t+1}^2) and the benchmark (u_{t+1}^2) for a 1–step–ahead forecast, the null-hypothesis of the *GW* test is $E\left[h_t \Delta L_{m,t+1} | F_t\right] = 0$, where F_t is an information set available at time t. Then, the test statistic is:

$$GW_{m,n}^1 = n\left(n^{-1}\sum_{t=T_1}^{T_2-1} h_t \Delta L_{m,t+1}\right)' \hat{\Omega}_n^{-1}\left(n^{-1}\sum_{t=T_1}^{T_2-1} h_t \Delta L_{m,t+1}\right) \overset{\alpha}{\sim} \chi_q^2, \quad (4.5)$$

where n=(T_2-1)-(T_1-1)=T_2-T_1 is the number of out–of–sample forecasts, m is the estimation sample, $\hat{\Omega}_n = n^{-1}\sum_{t=T_1}^{T_2-1}\left(h_t \Delta L_{m,t+1}\right) \times \left(h_t \Delta L_{m,t+1}\right)'$ is a consistent variance estimator and the test function h_t is set to $[1\ \Delta L_{m,t}]$.[25]

To test the significance of our survey indicators before and during the recession, we split the sample 11/2006–09/2009 into a pre–crisis and a crisis period, where the latter starts from 11/2008.[26] This choice is based on the dramatic

[25] Setting the conditioning set in this way, we expect past loss differences to predict their future dynamics, i.e. to indicate the more accurate model (Giacomini and White 2006).

[26] Alternatively, we also try different starting periods shifting 11/2008 by one month back and forth. The results remain robust and are available upon request.

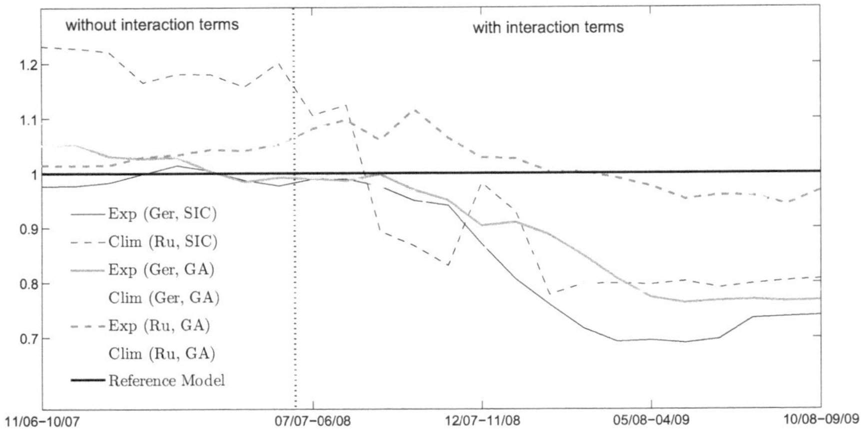

Figure 4.5: One–year forecast accuracy compared to AR with IEW (TU) via GA

We denote in our VAR models Germany with 'Ger' and Russia with 'Ru', while 'Exp' and 'Clim' stand for business expectations and business climate.

change in the forecasting accuracy of the VAR models in regard to NF observed for both standard and heuristic model selection strategies. To this end, we employ the IEW approach constructing 1–step–ahead forecasts for the entire pre–crisis ('prior') and crisis ('during') subsamples. Results of the pairwise test of equal conditional predictive ability are given in Table 4.4.

While for Germany constructed VAR models demonstrate significantly better forecasting performance only in the latter subsample (during the recession), resulting VAR models for Russia significantly (but negatively) differ from the benchmark in the former one (prior to the crisis). This finding demonstrates a marked difference between the results obtained via the heuristic strategy (right panel) and the '*all up to p-th lag*' method (left panel). Thus, employing the standard approach one could obtain evidence confirming either significantly superior or inferior performance of the VAR models during the recession depending on the BTS and IC chosen. However, since the heuristic approach consistently demonstrates better forecasting performance (Figures 4.4 and 4.5), its results can be considered as more justified alternative.

Differences in Data Publication

As it was mentioned earlier, BTSs are particularly useful because they are published more promtly than IP indices. However, once one is interested in real

Table 4.4: Conditional predictive ability test results[a]

Model specification			'all up to *p*-th lag' prior	'all up to *p*-th lag' during	GA prior	GA during
Germany	Business expectations	AIC	0.8810	0.8490	1.0010	0.7602*
		SIC	0.9530	0.7415		
	Business climate	AIC	0.8677	0.8034*	1.0210	0.7735**
		SIC	0.9788	0.8204		
Russia	Business expectations	AIC	1.1599	1.0241**	1.0634**	0.9541
		SIC	1.1755*	0.7989		
	Business climate	AIC	1.2335*	0.8791	1.1177*	0.9061
		SIC	1.1714*	0.8021*		

[a]The entries are the relative *RMSFE*s of corresponding VAR models with respect to AR benchmark with **,* denoting significance of the Giacomini and White (2006) test at the 5 and 10% level, respectively.

time forecasts, it is important to control for publication delay of the time series used. In our study, IP indices are published with a delay of 35 and 15 days for Germany and Russia, respectively. BTSs in the two countries, in contrast, are published more synchronously (at the end of the respective month). This implies for the models we build (see 4.1) that the first lag of IP (Y_{t-1}) for Germany is not available for the time the forecast is made (for Russia no additional restriction in data availability is applicable). To account for this important difference we construct VAR and AR models using standard and GA strategies and excluding Y_{t-1}, and repeat the exercise made before. Results are dispayed in Figure 4.6 and Table 4.5.

Table 4.5: Conditional predictive ability test results accounting for the restriction in data availability[a]

Model specification			'all up to *p*-th lag' prior	'all up to *p*-th lag' during	GA prior	GA during
Germany	Business expectations	AIC	0.8803	0.7806*	1.0000	0.8584*
		SIC	0.9368	0.8262		
	Business climate	AIC	0.8441	0.8094	1.0283	1.0164
		SIC	0.9753	0.8653		

[a]The entries are the relative *RMSFE*s of corresponding VAR models with respect to AR benchmark with **,* denoting significance of the Giacomini and White (2006) test at the 5 and 10% level, respectively.

Clearly, the supremacy of the VAR models over the AR benchmark becomes less stable when imposing the restriction that the first IP lag should not be included (GA, however, still consistently outperforms the standard model selection strategy). For example, shifting the out-of-sample period by just one month

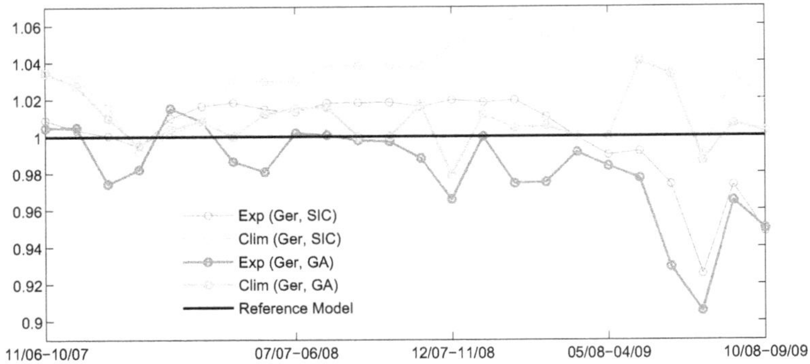

Figure 4.6: One–year forecast accuracy compared to AR with IEW (TU) via GA accounting for the restriction in data availability

may shift the results in favour of either an AR or a VAR model (Figure 4.6). Hence, the GW test provides less support for the contribution of the German BTSs to the forecasting performance (results only significant at the 10% level for business expectations and not significant for business climate). This finding can be explained by a fairly unstable contribution of the first IP lag to forecasting performance in such a volatile period as the recent crisis. Thus, comparing Figures 4.5 and 4.6 we see that depending on a particular out-of-sample period exclusion of this information improves the AR benchmark's relative performance to a certain extent, sometimes making it even a dominant strategy.

This latter finding provides an important argument in favour of an earlier publication of real economic data by statistical offices (as Destatis in our case) as this is crucial for the quality of the forecasts to be produced. In particular, while Russian BTS do not significantly contribute to the VAR models' performance, for Germany the evidence is dependent on the data availability issue: in the scenario when the IP value lagged one period is not available for Germany the improvement is marginal, while if all data would have become available at the time of the forecast, resulting forecasts could be improved significantly.

4.4.3 Dynamic Forecast

To provide an additional analysis of the two model selection strategies, we also consider a dynamic h-step-ahead forecast.

Consider first the standard model selection approach. We construct dynamic forecasts for the entire year 2009 (01/2009–12/2009) using the rest of the sample for estimation. The resulting forecasts for all three alternatives (AR models, VAR models with either business expectations or with business climate) via SIC are presented in Figure 4.7 (left panel for Germany and right one for Russia).

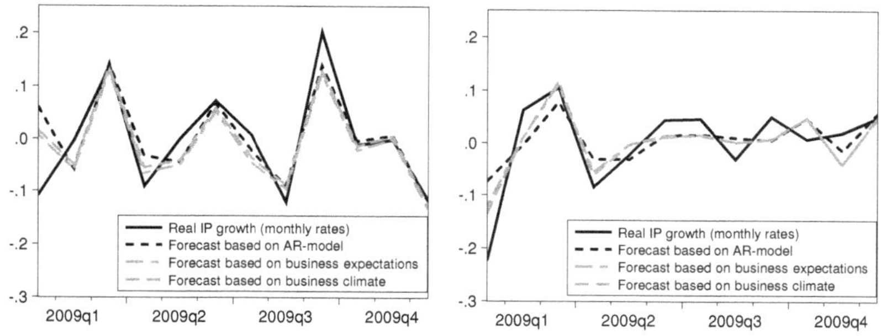

Figure 4.7: Prediction dynamics for the IP growth rate for Germany and Russia

The resulting VAR forecasts have minor differences predicting fairly similar IP growth dynamics. Estimating yearly IP growth rates (calculated as average of monthly year-to-year growth rates)[27] using the best unbiased model specification alternatives according to Table 4.1 (VAR with business expectations via SIC for both countries), one obtains results as in Table 4.6.

Both of the VAR–forecasts outperform the predictions of the corresponding benchmarks (AR and NF) in the two countries. It is also clear that the VAR–forecasted IP growth rate is much closer to the actual rate for Germany than for Russia. The large difference between the forecasted and the actual rates in the latter case might be due to the shorter expectation horizon of Russian BTSs. Thus, while the survey indicators in Russia can be accurate enough in the one–step–ahead forecast (see Figure 4.2), they are obviously inferior for the multi–step–ahead approach.

In contrast, employing the subset models obtained via GA for AR and VAR (with business expectations) models, forecasts constructed for IP in Russia exhibits a slightly improved approximation of the actual growth rates. This holds for both AR and VAR models, whereas their difference is very small. The latter finding is consistent with the *GW* test results. For Germany, in contrast, GA do not make any difference for the AR model yielding the same prediction. The VAR forecast, however, is also slightly improved in comparison to the standard strategy.

[27]This methodology is also used by Thompson Reuters.

Table 4.6: Prediction accuracy for the IP growth rate in 2009

		NF forecasts	AR forecasts	VAR forecasts	Actual rates
Germany	'all up to p-th lag'	-14.45%	−6.13%	−17.51%	-15.99%
	GA		−6.13%	−16.59%	
Russia	'all up to p-th lag'	+0.11%	−4.16%	−5.68%	-10.84%
	GA		−7.52%	−7.42%	

4.5 Conclusion and Outlook

While there are many studies providing evidence that leading indicators improve univariate time series models forecasting of real output in OECD countries, this is by far not the case for developing countries like Russia. Moreover, there is a lack of research examining the indicators' performance during the crisis. Therefore, our study contributes to the existing literature in two ways: by analyzing business tendency surveys (BTSs) with regard to their performance in predicting IP in Russia and Germany both prior and during the recent economic recession.

For this purpose, we specify VAR models accounting for a structural break and seasonal variation in the BTSs. Due to the limited data sample available and the large number of parameters to be estimated, high variance in estimation errors and, hence, resulting forecasts is obtained. To reduce this problem, an alternative subset selection strategy (GA) allowing for 'holes' in the lag structures is implemented. Since the data set is too small for IC to identify the 'correct' model, *RMSFE* for a pre-testing sample is taken as an objective criterion. The heuristic strategy is assessed in its in–sample and out–of–sample fit demonstrating superior forecasting performance.

Constructing 1–step–ahead forecasts for rolling windows of one and two years length with both increasing and fixed estimation windows, we demonstrate that while for Russia there is no significant difference between VAR and AR forecasts during the crisis (prior the crisis the difference is significant, but negative), for Germany the evidence is dependent on the data availability issue: in the scenario taking the restricted data availability into account, the improvement is limited, while if all data would have become available at the time of the forecast, resulting forecasts could be improved significantly. These results are confirmed by the conditional predictive ability test. The latter fact provides an important argument in favour of an earlier publication of real economic data by statistical offices (as Destatis in our case) as this is crucial for the quality of the forecasts to be produced.

Also, we show that these results can be obtained employing GA, but not the standard '*all up to p-th lag*' strategy. This seems to be a result of the very limited data sample, where a comprehensive model selection strategy is required to optimize multivariate lag structures sparing some degrees of freedom and, hence, reducing variance in resulting forecasts.

In conclusion, we argue that the heuristic methods are an effective tool of model selection also for short time series models. As this problem is not extraneous for Germany either (due to, e.g., its reunification) and considering the fact that this model selection technique has not been applied in the context for Germany before, the study is seen to be informative for both countries. For future study remains their comparison with other alternatives as, e.g., Bayesian model averaging or shrinkage estimators. Furthermore, in a few years it might be worth to repeat the experiment on a larger data set (for Russia) to test whether the BTSs would provide any better results.

Acknowledgements Thanks are due to participants of the 5th ifo workshop on macroeconomics and business cycles in Dresden and of the 3rd international conference of the ERCIM in London for their constructive comments. We also thank two anonymous referees for suggestions that led to substantial improvements in the manuscript. All shortcomings are our responsibility.

Financial support from the German Academic Exchange Service (DAAD), the EU Commission through MRTN-CT-2006-034270 COMISEF and the German Science Foundation (DFG RTG 1411) is gratefully acknowledged.

4.6 Appendix

4.6.1 Tests of Nonstationarity

To test for possible nonstationarity of the variables we apply the Augmented Dickey-Fuller-test (ADF–test) (Dickey and Fuller 1979) to variables, where no structural shift is assumed, and an UR–test (see Lanne *et al.* (2002)) to the BTSs of Russia with structural breaks (UR–test).[28] The null–hypothesis (H_0) of both, the ADF– and the UR–tests, is that the variable under consideration has a unit root. The alternative hypothesis is that it is stationary. For all series the test is run with a constant and the number of lags included to be selected based on SIC with a maximum lag order of 12 months. The test statistics are summarized in the lower left entries in Table 4.7.

[28] The latter test is implemented using the JMulTi free statistical software package available at http://www.jmulti.de/.

Table 4.7: Unit root test results

	Germany			Russia		
	lags	ADF–test	KPSS–test	lags	ADF/UR–test	KPSS–test
IP growth rate	12	1.76	0.24	11	−2.25	0.11
Business expectations	1	3.08**	0.18	1	−2.18	
Business climate	2	−2.47	0.13	0	−5.39***	

***/**/* denote significance at the 1, 5 and 10% level, respectively.

From a theoretical point of view it is expected that the IP growth rates and the BTSs are stationary. However, this assumption is not confirmed for IP growth rate[29] and two of the four survey indicators. These results might have been expected due to the low power of the test against H_0.

Starting with H_0 of stationary series and running the KPSS–test (Kwiatkowski *et al.* 1992) to the variables with no structural shifts, including again a constant and determining the bandwidth with the Newey–West procedure, results in no rejections of the hypothesis (see upper right entries in Table 4.7). As a result, we cannot reject the null–hypothesis for business expectations in Russia, but due to the low power of the ADF–test we presume the variable to be stationary, e.g., owing to its constraint on the interval $[-100; 100]$.

4.6.2 Lag Length Selection

After a specific model specification is selected, we test for possible residual autocorrelation. To this end, we employ the multivariate extension of the test suggested by Box and Pierce (1970), namely the Ljung–Box portmanteau test with the maximum lag length of 13 (Table 4.8) with its p–values in parentheses.

It becomes obvious that the models for the German data do not exhibit residual autocorrelation at the 5% significance level when choosing the lag length suggested by SIC. This also holds true for the Russian VAR model with the business expectations, but not for the business climate. This result might be due to changes in the seasonal pattern of the time series described in Section 4.2. A more detailed analysis of this feature is left for future research. For the current analysis we stick to the lag orders presented in Table 4.8. In contrast,

[29] We believe that the main reason for this is the extraordinary drop of the IP growth rates at the end of 2008 due to the global economic crisis. In fact, considering the shorter sample until 09/2008 supports the stationarity assumption (results available upon request).

Table 4.8: Lag length selection results

	SIC				AIC			
	Germany		Russia		Germany		Russia	
	lags	Test	lags	Test	lags	Test	lags	Test
Business expectations	2	58.2 (0.07)	1	62.8 (0.08)	5	45.0 (0.06)	3	53.5 (0.08)
Business climate	2	53.9 (0.15)	1	74.5 (0.01)	4	30.2 (0.74)	3	40.0 (0.52)

selecting the lag length according to AIC one has to deal with larger models (as expected), but identifies no autocorrelation at the 5% level.

4.6.3 Forecasts' Bias Analysis

Table 4.9: Forecasts' bias analysis

Model specification		$\alpha = 0$		$\beta = 1$		$\alpha = 0$ and $\beta = 1$	
		t-statistics				*p*-values	
		Germany	Russia	Germany	Russia	Germany	Russia
SIC	AR model	-1.63	-1.52	-0.49	1.06	0.19	0.30
	Business expectations	-0.60	-0.66	-0.27	0.75	0.81	0.63
	Business climate	-0.97	-1.13	-0.45	0.88	0.55	0.43
AIC	AR model	-0.84	-1.52	-0.94	0.75	0.44	0.34
	Business expectations	-1.31	-1.50	-0.45	0.20	0.37	0.35
	Business climate	-1.66	-1.22	0.03	0.45	0.28	0.49

4.6.4 Forecast Accuracy in Relation to NF

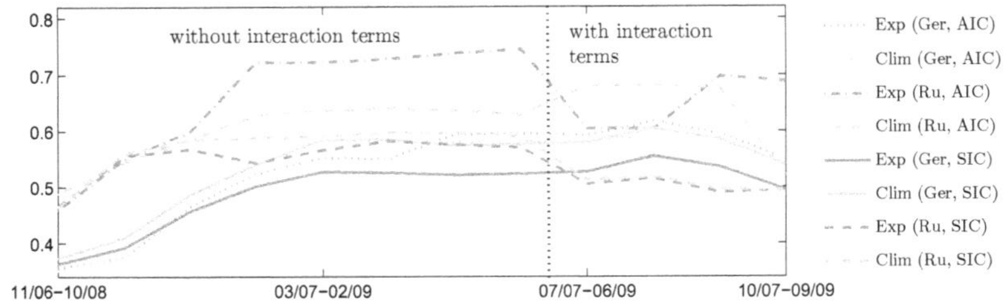

Figure 4.8: Two–year forecast accuracy in relation to NF with IEW

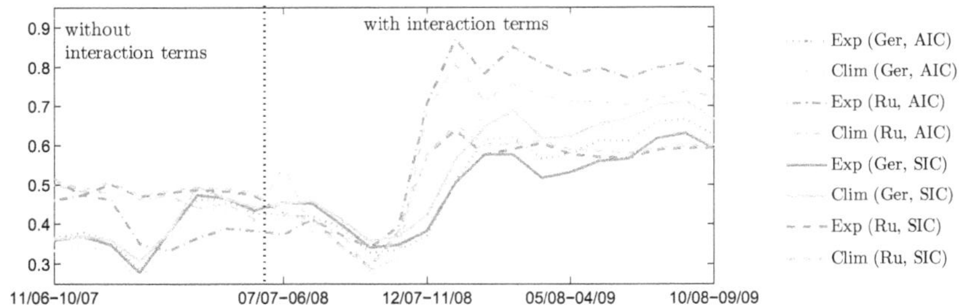

Figure 4.9: One–year forecast accuracy in relation to NF with FEW

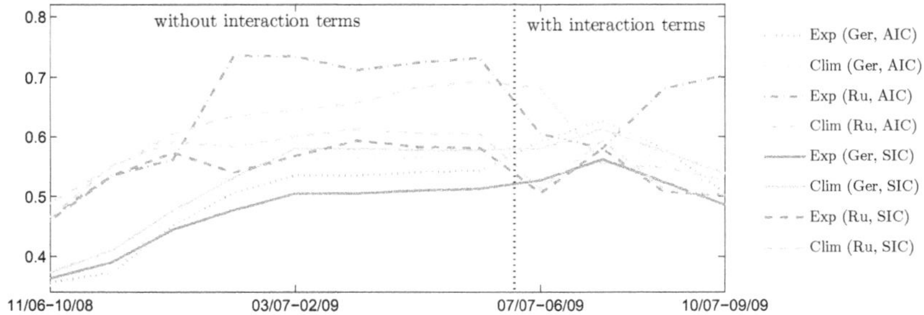

Figure 4.10: Two–year forecast accuracy in relation to NF with FEW

4.6.5 Forecast Accuracy in Relation to AR

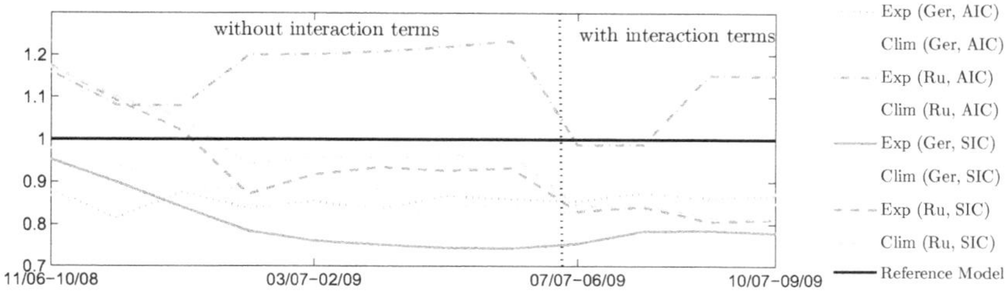

Figure 4.11: Two–year forecast accuracy in relation to AR with IEW

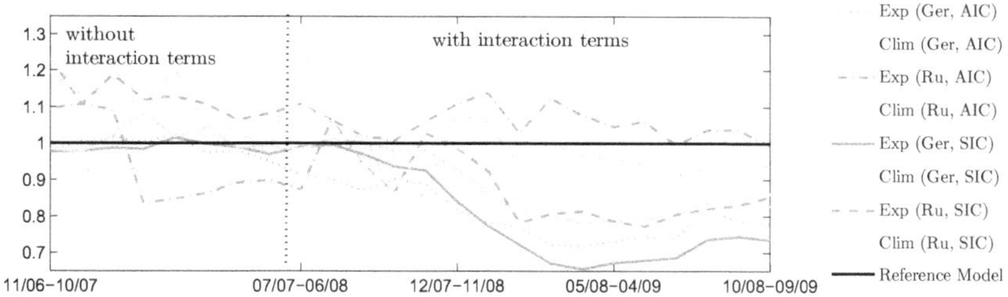

Figure 4.12: One–year forecast accuracy in relation to AR with FEW

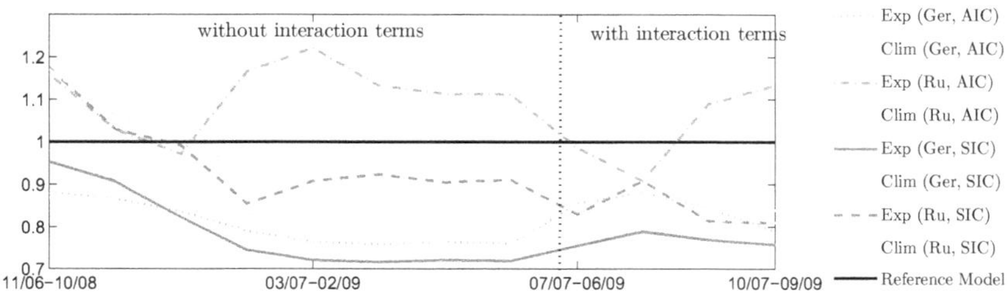

Figure 4.13: Two–year forecast accuracy in relation to AR with FEW

Chapter 5

A comparative study of the Lasso-type and heuristic methods

Ivan Savin
The final publication is available at http://www.jbnst.de/en/index.php

5.1 Introduction

The model selection process is crucial for the further analysis of any multiple regression model. Picking up too many regressors increases the variance of the constructed model, and taking fewer regressors than needed results in inconsistent estimates. In the last decade the least absolute shrinkage and selection operator (Lasso) (Tibshirani 1996) has become a very popular method for simultaneous model selection and parameter estimation.

Among the Lasso's main advantages are the combination of prediction accuracy and the parsimony of models built. The Lasso-type estimator outperforms simple application of parameter estimation methods (as, e.g., ordinary least squares or method of moments) since it shrinks the coefficients of insignificant regressors towards zero. Hence, the resulting models concentrate on the strongest effects and the total accuracy of the model forecast is increased. In addition, the Lasso solutions are more stable than other subset selection techniques based on the information criteria (IC) and stepwise strategies as, e.g., the general-to-specific approach (PcGets) discussed by Hendry and Krolzig (2005) and its bottom-up alternative (RETINA) analyzed by Perez-Amaral *et al.* (2003).

Another important advantage of the Lasso is its computational feasibility. Since its computational cost hardly exceeds the complexity of one linear regression (Efron *et al.* 2004), it is more attractive in comparison to classical model selection strategies that involve more intensive combinatorial search. However, the Lasso estimator has some limitations. In particular, inconsistent results are obtained for highly correlated regressors (see Section 5.2).

In the last years many studies have been devoted to methods revising and improving the initial Lasso concept. Since it is infeasible to describe them all in detail in this short introduction, I name only the most important ones from my perspective: the elastic net (EN) (Zou and Hastie 2005) and the adaptive Lasso (aLasso) (Zou 2006). A special case of the Lasso-type technique with the penalty term's exponent less than one is analyzed by Knight and Fu (2000).

This study compares the Lasso-type model selection strategies with one based on IC. In opposition to the Lasso, IC-based selection methods remain consistent even for data sets with correlated regressors. The IC's main constraint is the computational burden associated with the search for the optimum solution even for a moderate number of regressors. However, as is shown, e.g., by Maringer and Winker (2009), thanks to recent advances in heuristic optimization methods mimicking natural evolution processes, there are efficient algorithms able to select a model with at least a good approximation to the IC's global optimum. To the best of my knowledge, this article is the first that compares the Lasso-type and the heuristic model selection methods. An important contribution of this study is the demonstration that in certain situations (e.g., if regressors in a given data set are pairwise highly correlated or can explain the indicator of interest to a large extent) subset selection methods via heuristic algorithms can outperform the Lasso-type solutions.

The remainder of the paper proceeds as follows. Section 5.2 introduces both the Lasso-type methods and the heuristic model selection technique. Section 5.3 provides results of a Monte-Carlo analysis and Section 5.4 illustrates an application to a cross-country growth model. Finally, Section 5.5 concludes.

5.2 Model Selection Strategies

5.2.1 Least Absolute Shrinkage and Selection Operator

The least absolute shrinkage and selection operator (Lasso) was introduced by Tibshirani (1996). Initially suggested as a constrained version of the ordinary least squares, Lasso can be applied to a variety of estimation methods including, e.g., VAR-models (Hsu *et al.* 2007) and GMM-estimators (Caner 2009).

Numerous applications of this technique can be found in medicine, economics and other scientific fields (Foster *et al.* 2008, Hastie *et al.* 2009).

Consider the basic approach to the model selection problem for the following regression function:

$$y = \alpha + X^{opt}\beta + \varepsilon, \tag{5.1}$$

where α is an n-vector with all elements equal, X is an $n \times k$ matrix of k regressors and their values for n observations, β is a $k \times 1$ vector of their coefficients and ε is an $n \times 1$ vector of residuals.

In (5.1) X^{opt} refers either to the 'true' model in a Monte-Carlo simulation set-up or to an optimal approximation to the unknown real data generating process. Let us assume that the predictors have been standardized to have mean zero and unit length, and that the response has mean zero:

$$\sum_{i=1}^{n} y_i = 0, \ \sum_{i=1}^{n} x_{ij} = 0, \ \sum_{i=1}^{n} x_{ij}^2 = 1, \ i = 1,...,n, \ j = 1,...,k. \tag{5.2}$$

Hence, one can omit α without loss of generality. Then, the Lasso objective function can be presented as follows:

$$\widehat{\beta}_{Lasso} = \arg\min_{\beta} \left[\| y - X\beta \|_2^2 + \lambda \| \beta \|_1 \right]. \tag{5.3}$$

While the first term in the right part of equation (5.3) is just the residual sum of squares (RSS), the second term with $\lambda > 0$ is the amount of shrinkage the Lasso applies to the sum of the absolute values of the coefficients.[1] Hence, the Lasso can be referred to as a special case of the Bridge regression approach (Frank and Friedman 1993) imposing an upper bound on the L^q-norm of the parameters ($0 < q < \infty$) with $q = 1$:

$$\| \beta \|_q = \left[\sum_{j=1}^{k} |\beta_j|^q \right]^{1/q}. \tag{5.4}$$

Equivalently to (5.3), the Lasso chooses $\widehat{\beta}$ by minimizing RSS subject to a bound t on the L^1-norm of the parameters:

[1] λ is a tuning parameter to be defined using a data-driven method as, e.g., cross-validation (see below for more details).

$$\widehat{\beta}_{Lasso} = \arg\min_{\beta} \| y - X\beta \|_2^2 \quad \text{subject to} \quad \| \beta \|_1 \leq t, \text{with } t \propto 1/\lambda. \tag{5.5}$$

In the following the intuition behind the Lasso algorithm (a modification of the *LARS* algorithm) is briefly described.[2] One starts with an empty model (all coefficients are set to zero) and identifies the predictor x_ζ out of the full set of k regressors ($X_{n\times k} = (x_1, x_2, ..., x_k)$) most correlated with the response y:

$$\widehat{\zeta} = \arg\max_{\zeta} |\widehat{c}_\zeta|, \quad \text{where } \widehat{c}_\zeta = X'(y - \widehat{\mu}) \tag{5.6}$$

with $\widehat{\mu} = X_{\mathcal{A}}\widehat{\beta}$ being a prediction vector of regressors included in the model (respectively, one starts with $\widehat{\mu}_0 = 0$).

Transferring the ζ-regressor to the 'solution path' ($\mathcal{A}$) one needs to ensure that the next predictor x_ϑ to be included in $\mathcal{A}$ (x_ϑ is the most correlated covariate with the current residual) has as much correlation with $y - \widehat{\mu}_1$ as x_ζ. In other words, $y - \widehat{\mu}_1$ has to 'bisect' the angle between x_ζ and x_ϑ, so that $c_\zeta(\widehat{\mu}_1) = c_\vartheta(\widehat{\mu}_1)$. To this end, one increases $\widehat{\mu}_0$ in the direction of x_ζ:

$$\widehat{\mu}_1 = \widehat{\mu}_0 + \widehat{\gamma}_\zeta x_\zeta. \tag{5.7}$$

In the *LARS* algorithm $\widehat{\gamma}_\zeta$ is taken as the smallest positive value, so that another regressor can be included in the path fulfilling the condition of 'equally correlated regressors'. Starting from the third predictor one employs an 'equiangular vector' ($u_{\mathcal{A}}$) in (5.7) instead of the previous included regressor. The 'equiangular vector' is a unit vector constructed based on all covariates already transferred to $\mathcal{A}$ ($X_{\mathcal{A}}$) generating equal angles with the regressors.

In addition, the algorithm enforces that in each step $\psi = 1, 2, ..., \Psi$ (when a new regressor is included in $\mathcal{A}$) the sign of all predictors' estimates ($s_{\mathcal{A}}$) in the Lasso solution $\widehat{\beta}$ must agree with the sign of the current correlation[3] $\widehat{c}_{\psi,\mathcal{A}} = X'(y - \widehat{\mu}_{\psi-1})$:

$$s_{\mathcal{A}} = \text{sign}(\widehat{c}_{\mathcal{A}}) = \text{sign}(\widehat{\beta}_{\mathcal{A}}). \tag{5.8}$$

If the restriction (5.8) is violated, the corresponding regressor x_ρ is removed from $\mathcal{A}$ and, therefore, is removed from the calculation of the next equiangular direction ($u_{\mathcal{A}}$). However, later x_ρ can be re-included in $\mathcal{A}$, but the order of

[2] For a detailed (technical) explanation of all steps see Efron *et al.* (2004).

[3] Ψ is an additional stopping criteria limiting the maximum number of steps. In the *LARS* algorithm $\Psi = 8k$ that is usually enough for all k to be included in $\mathcal{A}$.

predictors in the solution path will be already different. This process continues until all k regressors are transferred to $\mathcal{A}$, thus, ensuring that at each step ψ only one regressor can be included or excluded from $\mathcal{A}$, $\max(\psi) \geq k$.

As a result, one obtains a piecewise-linear solution path in the tuning parameter $\lambda \in [0, \infty)$ with all $\widehat{\beta}$'s set to zero at $\lambda = \infty$ and equal to the OLS estimate at $\lambda = 0$ (all k covariates included).

Then, to select a single solution out of $\mathcal{A}$, λ is chosen by tenfold cross-validation producing a nearly unbiased estimate for the prediction error (PE_{CV}) of the model:

$$PE_{CV} = \frac{1}{10} \sum_{\varpi=1}^{10} \left(y_{\varpi} - X_{(\varpi)} \widehat{\beta_{\varpi}} \right)^2 . \tag{5.9}$$

In (5.9) the original sample is randomly partitioned into ten subsamples ($\cdot_{\varpi}$), whereas nine of them are used as training data to obtain $\widehat{\beta_{\varpi}}$ and a single one is retained as validation data for testing the model. The process is repeated ten times and the results from the folds are averaged. Alternatively, bootstrap re-sampling or the Stein's unbiased estimate of risk can be used (Tibshirani 1996).

The pseudocode of the procedure described is provided in Algorithm 3.

Algorithm 3 Pseudocode for the Lasso.

Generate an empty solution β_0, initialize k, Ψ, $\mathcal{A}$ and X
while $I \neq \emptyset$ and $\psi \leq \Psi$ **do**
    Select $\max|\widehat{c}_{\zeta}|$, transfer x_{ζ} into $\mathcal{A}$
    Identify $\widehat{\mu}$ that $c_{\zeta}(\widehat{\mu}) = c_{\vartheta}(\widehat{\mu})$
    for $x_{\rho} \subset \mathcal{A}$ **do**
        Estimate $\widehat{c}_{\mathcal{A}}$ and $\widehat{\beta}_{\mathcal{A}}$
        if $\text{sign}(\widehat{c}_{\mathcal{A}}) \neq \text{sign}(\widehat{\beta}_{\mathcal{A}})$ **then**
            Remove x_{ρ} from $\mathcal{A}$
        end if
    end for
end while
Select $\widehat{\beta}_{Lasso}$ with $\min(PE_{CV})$

In the following this study concentrates on the modification of the *LARS* method to solve (5.3). Thereby, the popular algorithm suggested by Efron *et al.* (2004) (and its modifications described below) is used for computational purposes.[4] However, one must admit that this is not the only technique available. The first algorithm was suggested by Tibshirani (1996), but turned out to

[4]Related codes available at http://www.stanford.edu/~hastie/Papers/LARS for R and http://www2.imm.dtu.dk/pubdb/views/publication_details.php?id=3897 for MATLAB.

be inefficient for large data sets (particularly large k). The first approach treating (5.3) as a convex problem was proposed by Osborne *et al.* (2000) deriving Lasso's dual problem and implementing a method similar to the one from Efron *et al.* (2004). Further methods include Turlach *et al.* (2005) (also via quadratic programming), Friedman *et al.* (2007) (via coordinate-wise optimization) and Figueiredo *et al.* (2007) (via gradient projection).

Setting $\lambda > 0$ by cross-validation one insures the Lasso solution to have a parsimony property, i.e. only a subset of resulting predictors in (5.3) has non-zero coefficients. This feature of the Lasso technique increases the total accuracy of the model forecast and makes the selected model easier to interpret.

However, the Lasso has substantial limitations. First, it is inconsistent when $k \gg n$ (underdetermined linear system). In this case, the Lasso can identify not more than $n-1$ (standardized) predictors (Efron *et al.* 2004). Second, it is also not able to identify all 'true' predictors in a data set with pairwise highly correlated regressors (Zou and Hastie 2005). The latter can be referred to as the 'irrepresentable condition' (Zhao and Yu 2006, p. 2544). As a result of the two constraints, the Lasso estimations can be biased.

Let us assume that in the 'true' model $\beta^{true} = \{\beta_1, \ldots, \beta_r, \beta_{r+1}, \ldots, \beta_k\}$ all non-zero coefficients are located between 1 and r. Then the matrix $C = \frac{1}{n} X_n^{'} X_n$ can be expressed in a block-wise form:

$$C = \begin{pmatrix} C_{11} & C_{12} \\ C_{21} & C_{22} \end{pmatrix} \text{ with } C_{11} \text{ being an } r \times r \text{ matrix.} \tag{5.10}$$

For the Lasso to be consistent, it is essential that

$$|C_{21} C_{11}^{-1} s| < 1 \tag{5.11}$$

with $s = (\text{sign}(\beta_1), .., \text{sign}(\beta_r))^{'}$ and 1 is a $(k-r) \times 1$ vector of ones so that the inequality (5.11) holds element-wise.

Thus, none of the irrelevant variables (the amount of its covariate) can be represented by the covariates of 'true' predictors. Otherwise the L^q-norm constraint on the regression coefficients has to be smaller than 1 ($q < 1$). (5.11) is known as the (weak) irrepresentable condition. It is always satisfied, e.g., for $k = 2$ or for the orthogonal design (uncorrelated regressors). More details on situations where the irrepresentable condition holds, can be found in literature (Zhao and Yu 2006, p. 2548).

Simply saying, Lasso is consistent in a low correlation setting:

$$\max_{j>r} \| \text{cov}(X_j, X^{true}) \text{cov}(X^{true})^{-1} \|_1 < 1, \tag{5.12}$$

while in presence of high correlations between 'true' and irrelevant variables, the Lasso cannot recover the correct sparsity pattern[5] ($\widehat{\beta} \nrightarrow \beta^{true}$). A detailed review of the literature on the situations, where Lasso can/cannot be consistent can be found in Yuan and Lin (2007, p. 145) and Gasso *et al.* (2009, p. 4687).

However, as Meinshausen and Yu (2008) show, even failing to discover the correct sparsity pattern (for (5.11) not being true) Lasso can provide good approximations of the 'true' model for large sample sizes ($\| \beta - \widehat{\beta}_{Lasso} \|_2 \to 0$ as $n \to \infty$). In other words, Lasso selects 'true' variables with high probability and irrelevant ones have only marginal coefficients (L^2-norm consistency).

5.2.2 Lasso Modifications

In the last years a large amount of studies has been devoted to methods revising and improving the initial Lasso concept. For this reason, the concepts discussed below are only a small sample from all the different modifications which have been proposed. For more complete overviews, the interested reader is referred to Hastie *et al.* (2009) and Bach and Obozinski (2010), among others.

In the following I concentrate on two main extensions of the Lasso: the elastic net (EN), which uses a combination of the Lasso and ridge regression penalty (Zou and Hastie 2005), and the adaptive Lasso (aLasso) applying different amounts of shrinkage for each regression coefficient (Zou 2006). The reason for this choice is twofold. First, the selected extensions are particularly designed to deal with the limitations stated above. Second, the two methods operate in a continuous space remaining computationally efficient.

Elastic Net

In many fields of application it is still common that only a small number of reliable observations exists for a large series of potential predictors ($k \gg n$). Numerous examples of this problem can be found in genetic engineering (e.g., gene expression data) or in chemometrics (e.g., fluorescence spectra) (Frank and Friedman 1993). In addition to the lack of degrees of freedom, these models include a set of highly correlated predictors. The latter problem can be encountered even for independent regressors X_k as long as $k \gg n$ (see Fan and Lv (2008, p. 852)).

In this case the standard Lasso algorithm is not the first choice (see Section 5.2.1). In order to overcome the problems described, EN includes an additional L^2-norm (ridge) shrinkage parameter into the objective function (5.3):

[5] According to Yuan and Lin (2007) inequality (5.12) is a sufficient condition for (5.11).

$$\widehat{\beta} = \underset{\beta}{\arg\min}\left[\| y - X\beta \|_2^2 + \lambda_1 \| \beta \|_1 + \lambda_2 \| \beta \|_2^2\right], \text{ with } \lambda_2 > 0. \tag{5.13}$$

Thanks to the added parameter in (5.13), the total EN penalty is strictly convex and, therefore, EN regression coefficients tend to be equal for highly correlated predictors, whereas the Lasso assigns two different (biased) coefficients (Zou and Hastie 2005).

To solve problem (5.13), one augments the original data set (y, X) of size $k+1 \times n$:

$$y^* = \begin{pmatrix} y \\ 0 \end{pmatrix}, \quad X^* = (1+\lambda_2)^{-\frac{1}{2}} \begin{pmatrix} X \\ \sqrt{\lambda_2} I \end{pmatrix}. \tag{5.14}$$

Due to the transformation in (5.14), the new data set (y^*, X^*) has the sample size $k+n$. Hence, EN can potentially select all k regressors.

Then the EN solution has the following form:

$$\widehat{\beta}_{EN} = \sqrt{1+\lambda_2}\widehat{\beta}^*, \text{ where } \widehat{\beta}^* = \underset{\beta^*}{\arg\min}\left[\| y^* - X^*\beta^* \|_2^2 + \frac{\lambda_1}{\sqrt{1+\lambda_2}} \| \beta^* \|_1\right]. \tag{5.15}$$

Thereafter, one takes a grid of values for $\lambda_2 = \{0, 0.01, 0.1, 1, 10, 100\}$ and performs the *LARS-EN* algorithm (as it is recommended in Zou and Hastie (2005)) for each of the values, selecting the one with the smallest PE_{CV}.[6]

Adaptive Lasso

Another approach 'correcting' the Lasso was introduced by Zou (2006) differentiating the amount of shrinkage for the coefficients. For this a vector of weights $(\widehat{\omega})$ is included in (5.3):

$$\widehat{\beta}_{aLasso} = \underset{\beta}{\arg\min}\left[\| y - X\beta \|_2^2 + \lambda \sum_{j=1}^{k} \widehat{\omega}_j |\beta_j|\right], \tag{5.16}$$

where the weights can be determined either by the OLS regression, $\widehat{\omega}_j = |\widehat{\beta}_{OLS}|^{-\upsilon}$ (if no collinearity is assumed), or by the ridge regression, $\widehat{\omega}_j = |\widehat{\beta}_{ridge}|^{-\upsilon}$ with $\upsilon > 0$. In the following only the 'ridge-weights' are used since they are more

[6]The choice of the grid is rather ad hoc and has to cover a large interval with a sparse number of values so that the two-dimensional crossvalidation is not too computationally costly (Zou and Hastie 2005, p. 310).

stable in the case of correlated predictors. As recommended by Zou (2006), $\upsilon > 0$ can be selected from the grid of values $\{0.5, 1, 2\}$ using two-dimensional cross-validation (the second tuning parameter is λ).

The objective function in (5.16) can be easily integrated in the *LARS* algorithm by defining $x_j^{**} = x_j/\widehat{\omega_j}$ and $\beta_{aLasso} = \widehat{\beta}^{**}/\widehat{\omega}$ (Zou 2006):

$$\widehat{\beta}^{**} = \arg\min_{\beta} \left[\| y - X^{**}\beta \|_2^2 + \lambda \sum_{j=1}^{k} |\beta_j| \right]. \tag{5.17}$$

For $n \to \infty$, $\widehat{\omega}_j$'s of 'false'-predictors grow to infinity applying an additional shrinkage for respective coefficients and, therefore, fulfilling the condition (5.11).

However, similarly to Lasso, aLasso is not consistent for $k \gg n$. In addition, for moderate sample sizes aLasso may not be dealing efficiently with correlated predictors. To the best of my knowledge, there is a lack of numerical studies on the performance of aLasso under these circumstances (the only exception I am aware of is presented by Zou and Zhang (2009)).

Further Modifications

It is important to note that the modifications described are only examples of the most popular extensions of the Lasso. Further literature on the Lasso-type methods includes, among others, Bach (2008), Candès *et al.* (2008), Lv and Fan (2009) and Zhang (2010).

Thus, Candès *et al.* (2008) suggest an alternative (multi-iterative) algorithm for the following reweighted L^1-norm minimization problem:

$$\widehat{\beta}^l = \arg\min_{\beta} \sum_{j=1}^{k} \| \widehat{\omega}_j^l \beta_j \|_1 \quad \text{subject to } y = X\beta, \text{ with } \widehat{\omega}_j^l = \frac{1}{|\widehat{\beta}_j^{l-1}| + c},$$

where l is an iteration count and c is a strictly positive stabilization parameter. The resulting $\widehat{\omega}_j$ then enter (5.16). Similar approaches of iterative reweighting algorithms can be found in Lv and Fan (2009) and Zhang (2010). Bach (2008) suggests a resampling approach (bootstrapping) to generate replications out of the original dataset serving as a basis to distinguish between relevant ('always included' by Lasso) and irrelevant (included with probability below 1) variables.

These methods, however, do not present a universal remedy from the limitations described. Comparison of these methods by simulation is the subject of future research.

5.2.3 Heuristic Optimization Methods

As an alternative to the Lasso, model selection based on information criteria (IC) is considered in this study. IC rank different models according to their fitness taking into account a penalty for model complexity. Over the last years IC have become a standard instrument in model selection ranging from lag order selection in multivariate linear (VAR and VEC) and nonlinear (MS-VAR) autoregression models to selection between rival nonnested models (Winker 1995, Winker 2000, Kapetanios 2007).

Consider a vector τ of length k with ones and zeros corresponding to selected and not selected regressors. To rank these vectors the Bayesian IC (BIC) and the Hannan-Quinn IC (HQIC) are implemented. Both these criteria have a similar structure:

$$\mathrm{IC} = \ln(\| y - X\widehat{\beta} \|_2^2) + f(h,n), \tag{5.18}$$

where the second term in the right part is a penalty dependent on the number of parameters included (h) and on the sample size (n). In particular, $h\ln(n)/n$ and $2h\ln(\ln(n))/n$ are the BIC and HQIC penalties.

Imposing some weak assumptions on the model space (x_i and ε_i) according to the results of Sin and White (1996), it can be shown that the vector τ^i that minimizes the IC converges to τ^{true} with probability close to 1 as $n \to \infty$. But for this to be true, it is essential that the penalty term $f(h,n) \to 0$ as $n \to \infty$. In this sense, BIC and HQIC are consistent.[7]

In general, the IC in (5.18) can be described as a L^0-constraint penalizing not the coefficients' values, but only their number:

$$\widehat{\beta}_{IC} = \arg\min_{\beta} \left[\| y - X\beta \|_2^2 + \lambda \| \beta \|_0 \right]. \tag{5.19}$$

As noted by Zhao and Yu (2006, p. 2553), the solution of (5.19) is consistent even for data sets with correlated regressors since it fulfills (5.11). However, since the search space of candidate models in (5.19) is discrete, the objective function is not necessarily 'well-behaved' enough to guarantee a global optimal using standard gradient methods, as the Newton or quadratic hill-climbing techniques. In fact, Breiman (2001) demonstrates the so called 'Rashomon Effect', where different model specifications with very similar IC values provide different conclusions. Hence, quality and precision of econometric estimation is crucially dependent on detecting the global optimum of (5.18). The full enumeration of all solutions is only feasible for a small k. In the following Monte-

[7]Note that IC consistency is considered stronger than the L^2-norm consistency of the Lasso since, in contrast to Meinshausen and Yu (2008), the correct sparsity pattern can be identified.

Carlo set-ups (Section 5.3) the selection is made out of 50 and 100 variables. Since a full enumeration of solutions results in 2^k potential sub-models, this is infeasible even using efficient algorithms such as those proposed by Gatu *et al.* (2008).

In the last decade a large number of studies has been devoted to subset selection methods. Among them are a sequential bottom-up (top-down) inclusion (deletion) of individual regressors (Perez-Amaral *et al.* 2003, Hendry and Krolzig 2005) using diagnostic tests or out-of-sample performance measures to check the validity of steps made. Another option is to set certain model prior probabilities shrinking the parameter search space. The resulting model averaging strategies are discussed in Kapetanios *et al.* (2008). However, these methods investigate only a portion of all potential submodels, whereas there is no good reason to assume that the 'true' model (in this case, the global optimum in terms of the IC) is within the small subset covered.

To overcome a priori restrictions on the search space that might result in local optima, one can take advantage of optimization heuristics that mimic natural evolution processes. These methods are called 'heuristic' because of their stochastic nature. Thanks to recent advances in heuristic optimization methods, there are efficient algorithms able to select a model with at least a good approximation to the IC optimum. A formal study on the convergence of heuristic algorithms can be found in Maringer and Winker (2009). For an overview of these techniques as well as for guidelines of their implementation, see Gilli and Winker (2009). In Savin and Winker (2012) a similar subset selection problem was handled by two heuristics: Threshold Accepting and Genetic Algorithms (GA). Since GA provided slightly better results in terms of both CPU time and solution quality, only GA are considered in the following.

GA are population-based heuristics (for examples of their application see **?**) that operate on a set of solutions (population). Thus, GA investigate the search space in many directions simultaneously, performing jumps in the search space by means of crossover and mutation mechanisms. Hence, the probability of getting stuck into a local optimum is reduced. The members in the population (chromosomes) are represented as bit strings, in which each position (gene) has two possible values: 1 and 0. In each generation GA replace parts of a population with new chromosomes (children) aimed to represent better solutions for a given problem. The GA pseudocode is described in Algorithm 4.

K is a matrix of $p = 500$ initial solutions generated by random distribution of zeros and ones.[8] Thereafter, the population is sorted according to (5.18). Then, the 50% of the chromosomes with the best target values (parents, K') are

[8]This number is considered to be large enough to screen the search space and to allow for effective selection of the best solutions.

Algorithm 4 Pseudocode for Genetic Algorithms.

Generate initial population K of solutions, initialize G and C
for $g = 1$ to G **do**
  Sort chromosomes in K
  Select $K^{'} \subset K$ (parents), select $K^{*} \subset K$ (elitist)
  initialize $K^{''} = \emptyset$ (set of children)
  for $c = 1$ to C **do**
    Select individuals $x^{parent1}$ and $x^{parent2}$ at random from $K^{'}$
    Apply cross-over to $x^{parent1}$ and $x^{parent2}$ to produce x^{child}
    $K^{''} = K^{''} \cup x^{child}$
  end for
  $K = (K^{'}, K^{''})$
  Mutate $K \setminus K^{*}$ at 5 random points
end for

transferred to the new population and new chromosomes (children, $K^{''}$) are constructed by crossing them over. Generating children one allows parents with superior objective values to be selected more often (see Savin and Winker (2012)). In this implementation the uniform crossover mechanism is used. Hence, parents may be split not only at one particular gene, but at each gene. Evidence on advantages of the uniform crossover technique can be found in Savin and Winker (2012).

After a new population is formed ($K = (K^{'}, K^{''})$), mutation is applied at five random genes with a probability of 50%.[9] All chromosomes in K except the ten best (elitist) solutions and the 10 children generated from the elitist solutions by mutation (K^{*}) are mutated. This procedure is repeated for a given number of generations $G = 2000$.[10]

An illustration on the resulting IC distribution (for BIC) for 100 Monte-Carlo restarts ($n = 400$, $k = 50$ with five of them involved in generating an artificial response variable) is shown in Figure 5.1: in the upper left plot the cumulative distribution function $F(IC)$ for different G is given, whereas the other plots are histograms of IC values identified. Increasing G the distribution shifts left and becomes less dispersed (Gilli and Winker 2009, p. 98). Since GA are stochastic methods, the algorithm is restarted ten times selecting the best IC.[11]

Important to highlight on this point is the difference between the two model selection strategies in identifying the optimal subset of regressors. While GA

[9] By this, the speed of searching through alternative solutions is increased, which is found to be a good strategy in conjunction with the uniform crossover mechanism (Savin and Winker 2012).

[10] As an alternative, one could stop the algorithm when ten best solutions of a generation (in terms of IC) are repeated (see Acosta-Gonzalez and Fernandez-Rodriguez (2007)).

[11] The GA code is available upon request.

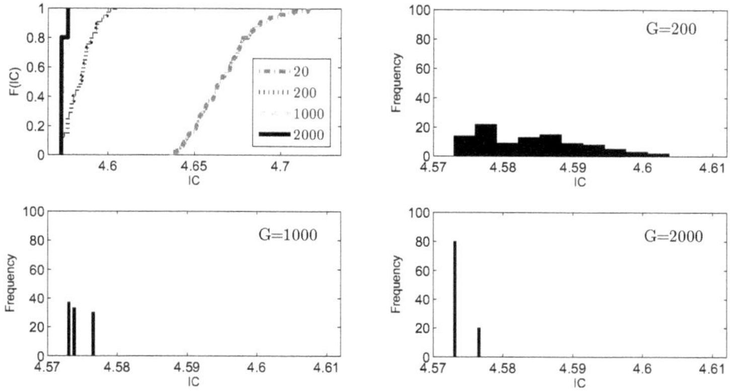

Figure 5.1: Empirical distribution of IC for different values of G.

minimize IC (highest model fit with a potentially moderate number of regressors), Lasso–type methods identify the subset with the smallest prediction error, which allows for parsimony of models built. Since this choice of penalties, however, is widely used and due to the fact that minimizing IC presents a highly complex integer optimization problem not compatible with continious Lasso algorithm, I compare the two strategies as they are presented above. In order not to mix estimation and model selection comparison of the two approaches, unregularized restricted estimation on the subsets obtained via all Lasso-type methods being tested is also employed.[12] Furthermore, since prediction error here is used as a penalty for model selection and is not necessarily meant for forecasting exercises, the two strategies (Lasso and heuristics tuned via IC) are compared based on their model selection and estimation quality, leaving the forecasting exercise out of this study. For details on the forecasting comparison of the two approaches, see Savin and Winker (forthcoming).

5.3 Monte-Carlo study

In this section the performance of the Lasso-type methods (Lasso, aLasso, EN) and the one of the subset selection technique via GA (BIC, HQIC) are compared. The goal is to determine how stable the methods are: in what set-ups

[12]This is in line with Candès and Tao (2007) arguing that unregularized restricted estimation alleviates a potential bias that results from the regularized estimation ($\lambda > 0$).

superior results are provided (in terms of correctly identified subsets and estimation accuracy) and what is the corresponding CPU time needed.

The Monte-Carlo set-ups below have certain parallels with the scenarios tested in Frank and Friedman (1993), Zou (2006) and Zou and Zhang (2009), making potential comparison of the results possible. However, there are also significant distinctions in the DGP (e.g., amount of noise, portion of relevant regressors) and in the scope of the methods tested (including also heuristics).

5.3.1 Data Generating Process

To compare the performance of the Lasso-type techniques with the IC based approach, various artificial data sets are generated. These set-ups are tested using different numbers of regressors in a data set and different numbers of observations per regressor ($k = 50$ or 100, $n = 60,\ 100,\ 400,\ 2000,\ 4000$ or 10000). Thus, the situations with $k \gg n$ and large n are covered.

The covariance matrix Σ is set either $\Sigma_{i,j} = 0.5^{|i-j|}$ or $0.75^{|i-j|}$ with $1 \leq i,j \leq k$. In the former case, all off-diagonal elements do not exceed 0.5 ('low correlation'); in the latter one, pairwise highly correlated regressors are generated ('high correlation'). The 'true' regression coefficient vector (β^{mc}) contains either a small or a large portion of non-zero coefficients ($k^{true} = 5$ or 25, respectively), which are either equal ($\beta^{mc}_j = 1$) or unequal ($\beta^{mc}_j = j^2$). In the latter case $\beta^{mc}_j = (1,4,9,16,...)$.

For each set-up, 50 restarts of the following procedure are performed. First, a set of regressors (X^{mc}) with a joint Gaussian distribution and a specified Σ is randomly generated. Then, using β^{mc} and adding an i.i.d. error term, the response variable (y^{mc}) is generated as follows:

$$y^{mc} = X^{mc}\beta^{mc} + \varepsilon, \quad \varepsilon \sim n(0,\sigma_\varepsilon^2), \tag{5.20}$$

where σ_ε^2 is the variance of the residuals.

In (5.20) one chooses σ_ε such that the corresponding signal-to-noise ratio (SNR) is either 5 ('low noise') or 0.5 ('high noise'), where the SNR is:

$$\mathrm{SNR} = \frac{\left(\mathrm{var}(X^{mc}\beta^{mc})\right)^{1/2}}{\sigma_\varepsilon}. \tag{5.21}$$

5.3.2 Simulation Results

The simulation results are compared using the True Positive Rate (TPR) and the False Negative Rate (FNR)[13] as statistics of a correctly identified model. The mean-squared error ($\text{MSE} = E[(\widehat{\beta} - \beta^{mc})'\Sigma(\widehat{\beta} - \beta^{mc})]$) with standard deviations computed over 50 replications given in parentheses are used as a measure of the estimation accuracy.[14] In addition, the CPU time corresponding to a single restart using MATLAB 7.11 on a Pentium IV 3.3 GHz is reported.[15]

As one can see in Table 5.1, Lasso-type solutions perform well identifying the correct subset structure in the scenario with low level of noise[16] (at most, only one false regressor included). It is also clear that in the case of high correlation and low noise level, EN outperforms other Lasso-type methods.

Table 5.1: Simulation results for $n = 400$, $k = 50$, $k^{true} = 5$ and $\beta_j^{mc} = 1$.

	Lasso	EN	aLasso	BIC	HQIC	Lasso	EN	aLasso	BIC	HQIC
	Low correlation					High correlation				
	Low noise									
TPR	98%	98%	81%	89%	67%	86%	91%	84%	91%	67%
FNR	0%	0%	0%	0%	0%	0%	0%	0%	0%	0%
MSE	.065	.058	.206	.005	.010	.069	.063	.461	.003	.010
	(.017)	(.015)	(.447)	(.004)	(.006)	(.020)	(.018)	(.858)	(.002)	(.005)
CPU	.1s	.7s	.7s	164s	177s	.1s	.7s	.7s	169s	182s
MSE_2	.002	.002	.007			.004	.003	.069		
	(.001)	(.001)	(.006)			(.003)	(.002)	(.333)		
	High noise									
TPR	94%	94%	97%	87%	65%	84%	84%	89%	80%	62%
FNR	5.8%	5.8%	4.1%	0.6%	0.2%	5.4%	5.3%	4.9%	1.7%	1.4%
MSE	4.67	4.61	3.57	.77	1.18	4.98	4.91	4.27	.90	1.26
	(.82)	(.89)	(1.19)	(.67)	(.67)	(.95)	(.90)	(1.12)	(.69)	(.66)
CPU	.1s	.7s	.6s	151s	169s	.1s	.7s	.7s	150s	165s
MSE_2	3.18	3.08	2.54			3.54	3.12	2.86		
	(1.28)	(1.28)	(1.32)			(1.35)	(1.33)	(1.28)		

However, in the scenario with high amount of noise, all Lasso-type methods tend to exclude two/three 'true' regressors from the solution identified (FNR≈4-6%). In this case aLasso performs best out of the other Lasso-type techniques.[17]

[13]TPR is the percentage of 'true' regressors from all variables selected and FNR is the portion of rejected 'true' regressors among correctly selected and correctly rejected ones. Thereby, ideally TPR=100% and FNR=0%. Consider an example: for a total of 50 variables with five of them being true, four are correctly selected, two are incorrectly selected and one is incorrectly rejected. Then TPR=4/6 ≈66% and FNR=1/(4+43)≈2%. Knowing k and k^{true}, one can think of actual number of variables incorrectly rejected/selected.

[14]Unregularized restricted estimations for Lasso-type methods are reported as MSE_2.

[15]For each method, averaged results over 50 replications of the procedure are reported.

[16]i.e., when a data set includes the majority of significant predictors of y.

[17]This fact is supported in other simulation studies (see, e.g., Johnson (2009, p. 496)).

If the regressors are also highly correlated, the probability of type I error (include false regressor) increases for all Lasso-type estimators. This results in a larger estimation bias.

In contrast, the results of the heuristic method are not influenced as strongly by the amount of noise in the simulated data sets. For small SNR IC via heuristics are better off identifying the 'true' subset. This is most obvious with the Bayesian IC that provides sparser models in comparison to HQIC (on average, BIC includes or excludes not more than one variable incorrectly).

Accordingly, IC tuned by GA clearly outperform Lasso-type methods in estimation accuracy in case of correlated regressors and/or high noise level. Notable, however, is that even in the case with low noise and low correlation (where Lasso and EN produce better subset recovery results) regularized Lasso-type estimations demonstrate a larger bias than heuristics. The bias is related to the soft threshold estimator property of Lasso (Tibshirani 1996, p. 285) shifting the values of large coefficients. Results of an unregularized but restricted estimation (on the subset selected) confirm this and alleviate the bias.[18]

In terms of the CPU time, the Lasso-type methods have a significant advantage over the heuristic approach taking not more than 1s to complete. Since EN and aLasso require a two-dimensional cross-validation, they need some more time than standard Lasso. IC via GA require about 170s per restart for $n = 400$. Reducing n results in a corresponding decline in the CPU time.

In Table 5.1 the high variance in the estimated bias (regularized estimation) for aLasso (especially for low noise level) is remarkable. In an additional experiment the described set-up is simulated 100 times for SNR $\in (0.3, 10)$ and the coefficient of variation for all three Lasso-type solutions is measured (Figure 5.2). Obviously, for SNR > 0.5 the variation in results for aLasso is much higher than for other methods. Similar results can be obtained for both ridge- and OLS-weights and are present in all the simulation studies performed.[19]

In the following one or two characteristics in the simulation set-up presented in Table 5.1 are changed and only major differences in results are reported. Thus, considering different regression coefficients (Table 5.2), the relative supremacy of the heuristic approach remains. All methods tested under this scenario exclude more correct regressors, which results in a higher MSE.

Increasing the portion of 'true' regressors ($k^{true} = 25$), one finds that for low noise heuristics outperform the Lasso-type methods in identifying the correct subset structure (Table 5.3). There can be two reasons for this. First, due to the stronger parsimony property of the Lasso methods (aLasso excludes cor-

[18] The finding is made throughout the study for $n \gg k$ or a data set with 'low noise' level.

[19] This is mainly due to the asymptotic property of the aLasso-weights ($\widehat{\omega}_j$) adding some instability in the model estimation. For SNR > 0.5 and, respectively, smaller MSE values this instability becomes more evident.

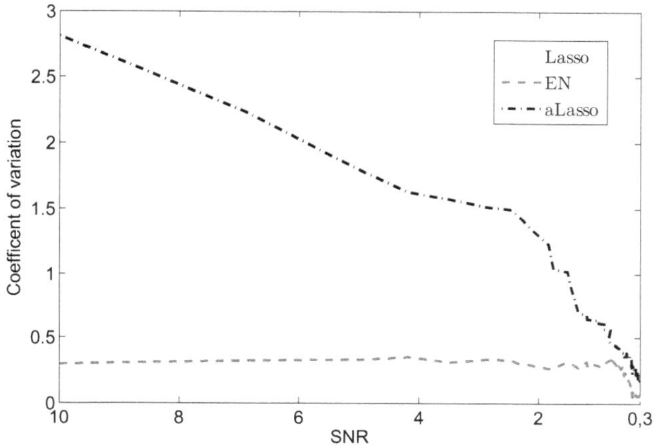

Figure 5.2: Coefficients of variation for Lasso-type estimators.

rect variables even with the small portion of noise). Second, due to the larger proportion of relevant predictors, the problem of correlated predictors is more challenging. This is evident when one compares the left and the right panels of Table 5.3. Increasing the amount of noise leads to a much larger proportion of mistakes in model selection for all methods. Nevertheless, the heuristics still results in a lower bias (both in terms of MSE and MSE_2).[20]

To see whether results of the methods improve for larger data sets (e.g., due to L^2-norm consistency defined by Meinshausen and Yu (2008)),[21] the simulations described in Tables 5.2-5.3 are repeated with n=2000 (see Tables 5.4-5.5).

As expected, the quality of both, subset selection and estimation, increases for heuristic and Lasso-type methods as n rises. However, the gain is larger for IC-based methods, particularly for the case of a sparse sample of 'true' regressors with large coefficients (Table 5.4). Thus, for 'high noise' the Lasso-type methods on average incorrectly exlude two variables, whereas IC - only one, which in conjunction with large coefficients determines a much larger bias.

In the case of a large k^{true} (Table 5.5), heuristics continue to outperform the Lasso-type methods for the 'low noise' set-up, but not for the 'high noise' one.

[20] Due to the higher k^{true} the heuristic requires more CPU time (approximately 230s) identifying all 'true' predictors under low noise and estimating their regression coefficients.

[21] Note that this study does not perform the additional step suggested by Meinshausen and Yu (2008) deleting variables with small absolute coefficients identified.

Table 5.2: Simulation results for $n = 400$, $k = 50$, $k^{true} = 5$ and $\beta_j^{mc} = j^2$.

	Lasso	EN	aLasso	BIC	HQIC	Lasso	EN	aLasso	BIC	HQIC
	Low correlation					High correlation				
	Low noise									
TPR	98%	98%	90%	92%	68%	87%	88%	90%	89%	66%
FNR	1.7%	1.7%	1.8%	.4%	.5%	1.7%	1.7%	2.1%	.7%	.6%
MSE	13.0	12.5	26.7	1.1	2.2	16.1	15.7	28.7	1.2	2.2
	(3.7)	(3.6)	(43.8)	(1.1)	(1.3)	(4.4)	(4.5)	(54.8)	(.7)	(1.3)
CPU	.1s	.6s	.7s	161s	174s	.1s	.7s	.7s	164s	177s
MSE_2	1.4	1.4	9.6			1.6	1.5	8.5		
	(.3)	(.3)	(19.7)			(.5)	(.5)	(15.6)		
	High noise									
TPR	100%	100%	92%	83%	65%	95%	96%	89%	79%	61%
FNR	6.6%	6.6%	5.5%	4.3%	3.7%	6.4%	6.3%	5.9%	4.0%	4.2%
MSE	847	823	466	123	178	899	858	614	124	157
	(168)	(169)	(333)	(74)	(80)	(187)	(181)	(351)	(81)	(84)
CPU	.1s	.7s	.7s	154s	171s	.1s	.7s	.7s	155s	172s
MSE_2	319	320	225			344	310	270		
	(127)	(127)	(142)			(143)	(138)	(136)		

Table 5.3: Simulation results for $n = 400$, $k = 50$, $k^{true} = 25$ and $\beta_j^{mc} = 1$.

	Lasso	EN	aLasso	BIC	HQIC	Lasso	EN	aLasso	BIC	HQIC
	Low correlation					High correlation				
	Low noise									
TPR	82%	84%	99%	98%	94%	74%	75%	99%	98%	94%
FNR	0%	0%	1.4%	0%	0%	0%	0%	4.4%	0%	0%
MSE	.62	.59	5.06	.13	.15	.63	.61	7.60	.13	.15
	(.15)	(.15)	(5.93)	(.05)	(.05)	(.15)	(.15)	(11.1)	(.05)	(.05)
CPU	.3s	1.0s	.9s	288s	297s	.3s	1.0s	.9s	289s	299s
MSE_2	.15	.15	0.89			0.16	0.15	2.35		
	(0.04)	(0.04)	(1.97)			(0.04)	(0.04)	(4.90)		
	High noise									
TPR	79%	79%	88%	82%	76%	64%	65%	82%	71%	69%
FNR	24%	24%	23%	21%	20%	21%	21%	23%	22%	21%
MSE	41.3	41.2	38.4	29.3	25.2	51.1	50.9	57.3	23.4	20.1
	(3.1)	(3.2)	(3.5)	(4.1)	(4.4)	(11.4)	(11.3)	(9.2)	(4.1)	(4.1)
CPU	.3s	.8s	.7s	166s	192s	.3s	.8s	.7s	170s	188s
MSE_2	40.8	40.8	38.7			26.5	26.0	43.4		
	(3.2)	(3.2)	(3.9)			(10.2)	(10.1)	(12.4)		

Table 5.4: Simulation results for $n = 2000$, $k = 50$, $k^{true} = 5$ and $\beta_j^{mc} = j^2$.

	Lasso	EN	aLasso	BIC	HQIC	Lasso	EN	aLasso	BIC	HQIC
	Low correlation					High correlation				
	Low noise									
TPR	100%	100%	91%	97%	76%	98%	98%	96%	96%	79%
FNR	.8%	.8%	1.5%	0%	0%	1.2%	1.1%	1.7%	.1%	0%
MSE	5.21	5.08	26.2	.14	.33	5.19	5.08	28.9	.16	.29
	(1.35)	(1.39)	(53.2)	(.12)	(.21)	(1.38)	(1.35)	(45.9)	(.16)	(.24)
CPU	.6s	1.3s	1.6s	673s	729s	.7s	1.5s	1.8s	688s	733s
MSE_2	.73	.51	8.8			.73	.65	10.9		
	(0.46)	(.49)	(19.9)			(.49)	(.48)	(24.5)		
	High noise									
TPR	100%	100%	95%	93%	69%	99%	99%	95%	89%	66%
FNR	5.1%	5.1%	4.2%	2.5%	2.3%	5.0%	4.8%	4.9%	2.9%	2.6%
MSE	380	371	224	19	31	382	375	318	20	33
	(92)	(92)	(239)	(11.5)	(16.3)	(93)	(93)	(241)	(13)	(17)
CPU	.8s	1.8s	1.8s	583s	645s	.8s	1.7s	1.9s	587s	656s
MSE_2	97	97	96			99	87	145		
	(63)	(61)	(122)			(63)	(71)	(156)		

Table 5.5: Simulation results for $n = 2000$, $k = 50$, $k^{true} = 25$ and $\beta_j^{mc} = 1$.

	Lasso	EN	aLasso	BIC	HQIC	Lasso	EN	aLasso	BIC	HQIC
	Low correlation					High correlation				
	Low noise									
TPR	89%	92%	99%	99%	96%	79%	80%	99%	99%	96%
FNR	0%	0%	.04%	0%	0%	0%	0%	2%	0%	0%
MSE	.21	.20	2.80	.023	.027	.22	.21	6.45	.024	.027
	(.05)	(.05)	(3.86)	(.007)	(.007)	(.05)	(.05)	(9.05)	(.007)	(.008)
CPU	.5s	1.3s	1.7s	1325s	1341s	.5s	1.3s	1.6s	1339s	1347s
MSE_2	.028	.027	.04			.032	.031	.74		
	(.007)	(.007)	(.10)			(.008)	(.008)	(1.79)		
	High noise									
TPR	88%	88%	95%	94%	89%	77%	77%	91%	88%	83%
FNR	5%	5%	13%	12%	8%	7%	7%	18%	17%	14%
MSE	14.70	14.70	19.20	7.94	5.66	14.94	14.90	22.65	7.77	6.30
	(2.26)	(2.24)	(4.78)	(1.12)	(.69)	(2.63)	(2.60)	(5.42)	(.81)	(.98)
CPU	.7s	1.6s	1.9s	994s	1175s	.6s	1.3s	1.8s	903s	1031s
MSE_2	4.35	4.35	11.12			3.86	3.86	11.74		
	(.89)	(.89)	(5.89)			(.25)	(.25)	(3.68)		

In the latter case, Lasso and EN better minimize the type II error (reject 'true' regressors: 2-4 instead of 6-8) resulting in a better estimation accuracy (only if the unregularized estimation is applied). These results serve as an illustration for the L^2-norm consistency and hold for some larger sample sizes (e.g., $n = 4000$). However, for $n = 10000$ Lasso, EN and IC via GA identify all relevant regressor (FNR=0%), whereas IC accept less false regressors and, hence, produce a smaller bias (results available upon request).

Thus, with increasing sample sizes GA via IC better improve their subset selection and estimation performance. However, to outperform the Lasso-type methods n must be large enough. This translates in more computational time required. However, one must remember that in real data applications the CPU time spent by GA can be reduced due to i) no need to produce replications as for the Monte-Carlo simulations; and ii) possibility easily to run MATLAB restarts in a distributed computing environment.

If one reduces n to 100, Lasso-type methods (except aLasso) are less affected by the asymptotic property than IC (Table 5.6). In the case of 'low noise', IC are still better off in terms of MSE (but not MSE_2), but for the 'high noise' IC generate a larger estimation bias.[22] In general, this is good evidence that Lasso is more suitable for small n (Hsu *et al.* 2007, p. 3649).

Table 5.6: Simulation results for $n = 100$, $k = 50$, $k^{true} = 5$ and $\beta_j^{mc} = 1$.

	Lasso	EN	aLasso	BIC	HQIC	Lasso	EN	aLasso	BIC	HQIC
	Low correlation					High correlation				
	Low noise									
TPR	87%	93%	88%	74%	49%	71%	74%	88%	74%	48%
FNR	0%	0%	0%	0%	0%	0%	0%	.03%	0%	0%
MSE	.19	.17	.43	.04	.08	.21	.18	.46	.04	.08
	(.08)	(.07)	(.53)	(.03)	(.04)	(.09)	(.08)	(.54)	(.04)	(.04)
CPU	.3s	1.2s	.8s	65s	72s	.3s	.8s	.8s	68s	69s
MSE_2	.02	.02	.03			0.02	0.02	.04		
	(.01)	(.01)	(.03)			(0.01)	(0.01)	(.14)		
	High noise									
TPR	73%	73%	58%	46%	33%	45%	46%	45%	39%	25%
FNR	7.0%	7.0%	6.4%	5.4%	4.5%	7.6%	7.6%	7.0%	5.9%	5.8%
MSE	5.04	4.93	4.76	7.26	9.37	5.66	5.51	4.80	6.23	9.02
	(.44)	(.51)	(1.06)	(3.64)	(4.33)	(.60)	(.60)	(1.74)	(3.34)	(3.96)
CPU	.2s	.6s	.7s	62s	65s	.3s	.8s	.7s	55s	56s
MSE_2	5.06	5.06	5.20			5.34	5.32	5.85		
	(1.35)	(1.35)	(1.81)			(1.00)	(0.98)	(2.98)		

[22] Note that in this set-up and for $k \geq n$ (see below) MSE_2 is not necessarily better than MSE as long as there is 'high noise' in the data set (and, respectively, worse subset identification by the Lasso-type methods) due to a low number of degrees of freedom left for the restricted estimation versus the feature of the regularized estimation to shrink coefficients of included 'false' regressors to negligible values (not for aLasso due to small n).

Finally, considering an underdetermined linear system one finds that the performance of all methods dramatically decreases (Table 5.7). This is more evident for the heuristic approach. The case with $k \gg n$ can result in extremely small RSS values if a large number of available regressors ($k \sim n$) is included. As the IC's natural logarithm goes to minus infinity, the penalty on model complexity remains in the same (former) order of magnitude. Hence, the resulting difference in IC 'compensates' incorrect variables to be included by GA.[23] An illustration of this effect is presented in Figure 5.3. In the left plot with $n = 400$, GA identifies a smaller IC value (dashed line) than the one attributed to the 'true' subset structure ('IC-true'). The smaller the sample size, the larger the difference between the two values. In the extreme case with $k \gg n$ (right plot) this difference becomes much more apparent.[24]

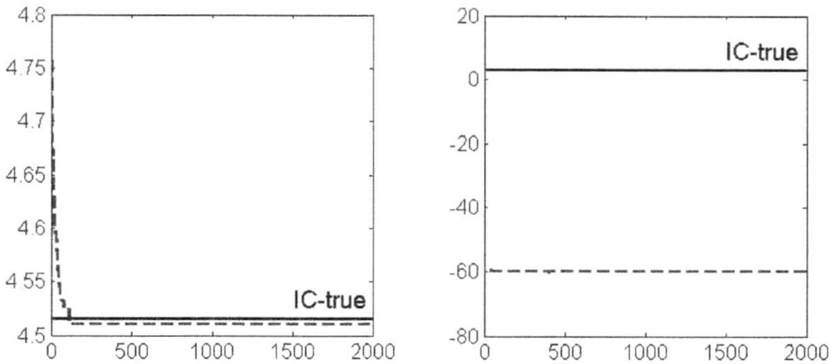

Figure 5.3: IC values for different sample sizes.

Consequently, the real limitation of the heuristic approach is the objective function (5.18) that is not suitable for $k \gg n$ in identifying correct regressors, while GA perform well in both set-ups minimizing the objective function. In the set-up with $k \gg n$, Lasso-type methods (in particular, Lasso and EN) are superior strategies in terms of both correctly identified subset structures and estimated bias.[25]

[23] A similar finding is also made for the Akaike information criterion.

[24] Simulation results for some additional experiments are provided at www.jbnst.de.

[25] Experiments with n=100 and k=500 or 1000 confirm the findings stated.

Table 5.7: Simulation results for $n = 60$, $k = 100$, $k^{true} = 5$ and $\beta_j^{mc} = 1$.

	Lasso	EN	aLasso	BIC	HQIC	Lasso	EN	aLasso	BIC	HQIC
	Low correlation					High correlation				
	Low noise									
TPR	52%	73%	44%	8%	8%	44%	52%	45%	8%	8%
FNR	0%	0%	0%	0%	0%	0%	0%	0.2%	0.2%	0.2%
MSE	.14	.14	.06	1.92	1.71	.16	.15	0.47	1.78	2.08
	(.08)	(.07)	(.06)	(.80)	(.59)	(.06)	(.08)	(1.17)	(.60)	(.78)
CPU	.4s	1.9s	1.9s	218s	225s	.5s	2.1s	1.6s	218s	224s
MSE_2	.09	.03	.14			.10	.09	.59		
	(.04)	(.02)	(.08)			(.06)	(.06)	(1.40)		
	High noise									
TPR	46%	46%	17%	5%	5%	35%	40%	16%	6%	6%
FNR	4.1%	4.1%	3.6%	3.8%	3.9%	4.3%	4.2%	3.6%	3.4%	3.4%
MSE	5.03	5.01	18.65	90.33	88.56	5.16	5.05	27.07	97.02	93.89
	(.22)	(.47)	(19.23)	(33.92)	(29.64)	(.41)	(.27)	(40.74)	(45.88)	(41.78)
CPU	.4s	1.7s	1.8s	222s	224s	.5s	1.9s	1.8s	224s	228s
MSE_2	6.38	6.38	26.43			6.79	6.53	41.05		
	(1.75)	(1.75)	(8.63)			(1.95)	(1.73)	(23.25)		

5.4 Application on a Cross-country Growth Model

To illustrate the model selection techniques, they are applied to an actual empirical problem, using the real per capita GDP growth rate over 1960-1992 and a series of country's characteristics evaluated for 1960.[26] The data set was first tested and described in detail by Sala-i-Martin (1997). Due to a large number of missing observations, Fernandez *et al.* (2001) reduce the original data set from 134 countries and 62 regressors to 72 and 42, respectively.

The data was used in a series of studies applying different model selection strategies. Most interesting for us (for comparative reasons) are the applications of GA by Acosta-Gonzalez and Fernandez-Rodriguez (2007) and aLasso by Schneider and Wagner (2009).

A brief summary of the results for the data set with the total number of regressors included (h) by each strategy is presented in Table 5.8 (see Table 5.9 in Appendix for a complete version).[27] As one can see, BIC via GA, Lasso and EN include more than half of the available regressors in the final solution. Comparing their model fits, the IC has the highest R^2 adjusted with the smaller subset of predictors.[28] This can be due to a larger number of incorrect variables included

[26]A set of characteristics is selected that best explains the GDP growth rate within a standard linear regression model.

[27]Note that for Lasso, EN and aLasso unregularized estimations and resulting (higher) R^2(adj.) are reported. Alternatively, the methods result in 78%, 81% and 67%, respectively.

[28]In contrast, aLasso rather excludes some relevant regressors resulting in the smaller R^2.

Table 5.8: Summary results for the cross-country growth model.

	BIC	BIC(adj.)	Lasso	EN	aLasso
Constant	0.0797	0.0419	0.0872	0.0800	0.1251
Primary school enrollment	0.0249	-	-	-	-
Life expectancy	0.0009	0.0010	0.0008	0.0008	0.0008
Log of per capita GDP	−0.0187	−0.0130	−0.0163	−0.0158	−0.0186
Fraction of confucius	0.0759	0.0577	0.0537	0.0592	0.0539
Fraction of muslim	0.0078	0.0118	0.0039	0.0024	-
Sub-Saharan dummy	−0.0218	-	−0.0204	−0.0184	−0.0259
Rule of law	0.0131	-	0.0099	0.0073	0.0199
Equipment investment	0.1511	0.2181	-	0.0917	-
...	...	...	...	...	...
	h=22	h=7	h=32	h=34	h=11
R^2(adjusted)	92%	79%	86%	87%	72%

by the Lasso-type methods for limited n (see results in Table 5.3). Based on this and together with our Monte-Carlo simulations, the model estimation obtained via GA is considered as the most accurate one in this particular example. Due to a potentially large portion of relevant regressors and high pairwise correlation between certain variables (e.g., for equipment investment and life expectancy it is above 0.64), EN is seen to outperform the other Lasso-type methods.[29] For more details on the pairwise correlation between the regressors see Figure 5.4 (variables selected by GA via BIC are grouped between 1 and 22; the main diagonal in the correlation matrix is removed). Clearly there is a high correlation (over $|0.5|$) not only within the grouped variables, but also between them and the rest of the data set.

Acosta-Gonzalez and Fernandez-Rodriguez (2007) try to avoid over-parametrization with an adjusted BIC by doubling its penalty on model complexity. They come up with a smaller model (see BIC(adj.)). By employing the increased penalty for GA, the same model as in Acosta-Gonzalez and Fernandez-Rodriguez (2007) is identified.

In Table 5.8 two out of three regressors always included by Sala-i-Martin (1997) are also selected by all the strategies tested: life expectancy and GDP per capita. In contrast, less evidence is for primary school enrollment to be retained in the model. This finding is also supported by Fernandez *et al.* (2001).

[29] In our application partly different results are obtained in comparison to the ones in Schneider and Wagner (2009). This is due to another choice of tuning parameters made. In particular, Schneider and Wagner (2009) employ OLS-weights and set υ equal to one.

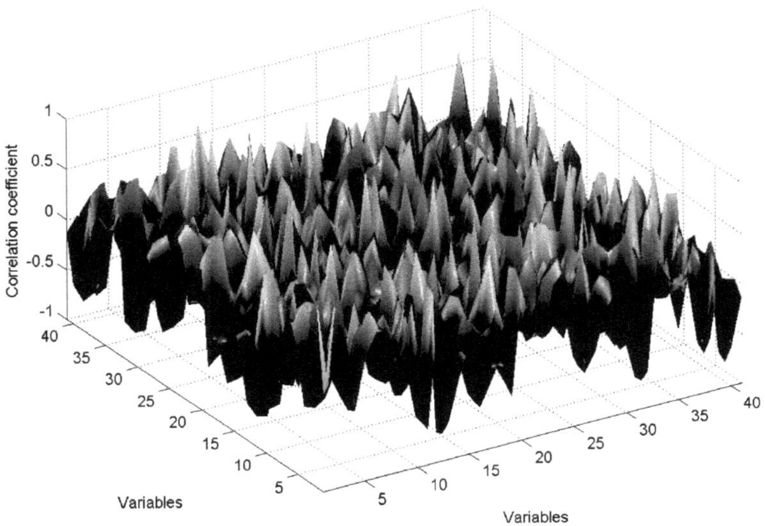

Figure 5.4: Pairwise correlation in the empirical data set.

5.5 Conclusions and Outlook

The model specification step has a vital role for further regression analysis, since any ad-hoc or intuitive decisions can reduce the estimation accuracy or introduce an estimation bias. In this study the Lasso-type (Lasso, aLasso, EN) and heuristic model selection strategies are compared. First, implementation of all methods is descibed underlining their strengths and weaknesses. Second, an illustration of their performances based on Monte-Carlo experiments and real empirical data is provided, and their results are contrasted.

One finds that an application of the Lasso modifications has some influence on its performance in terms of both subset selection correctness and estimation bias. However, this influence is rather marginal in comparison to other model selection methods, in particular, heuristic optimization. In contrast, a significant gain in estimation accuracy can be obtained via an additional unregularized restricted estimation as long as $n \gg k$ or the data set has a low level of noise (which is difficult to know *a priori*).

In general, the Lasso-type techniques provide sparser solutions than the heuristic approach. They can better identify irrelevant predictors in a final subset, but exclude more relevant ones. As a result, in most of the simulated set-ups the Lasso methods exhibit a larger bias (particularly, considering regularized estimations only). If regressors in a given data set are pairwise strongly correlated or can explain the indicator of interest to a large extent ('low noise'), the supremacy of the heuristics becomes more apparent.

Based on the simulated experiments, the Lasso methods are more suitable for data sets with small sample sizes (where heuristics are limited by asymptotic IC). In contrast, heuristics can perform much better on larger data sets.

The Lasso-type methods have a significant advantage over heuristic methods in terms of the CPU time required. Although, as it is demonstrated in this study, nowadays IC can be applied with reasonable computational effort

Future research might compare the methods discussed with the adaptive elastic net (Zou and Zhang 2009) that combines strengths of EN and aLasso, or with the adaptive ridge selector (Armagan and Zaretzki 2010) that differentiates shrinkage according to t-statistics. Furthermore, strategies from Section 5.2.2 can be also added to the comparison. A further extension is to test an application of heuristics on the generalized L^q-norm approach with $0 < q < 1$.

Acknowledgements

I thank Thomas Wagner, Marianna Lyra and Henning Fischer for their constructive suggestions. Special thanks are due to Peter Winker for his valuable advices with regard to heuristic optimization. I also thank three anonymous referees for suggestions led to improvements in the manuscript. All remaining shortcomings are my responsibility.

5.6 Appendix

Table 5.9: Results for the cross-country growth model.

	BIC	BIC(adj.)	Lasso	EN	aLasso
Constant	0.0797	0.0419	0.0872	0.0800	0.1251
Primary school enrollment	0.0249	-	-	-	-
Life expectancy	0.0009	0.0010	0.0008	0.0009	0.0008
Log of per capita GDP	-0.0187	-0.0130	-0.0163	-0.0158	-0.0186
Fraction of GDP in mining	0.0328	-	0.0484	0.0367	-
Degree of capitalism	-	-	0.0023	0.0018	-
Number of years open economy	-	0.0176	0.0006	-3.8×10^{-5}	-
Fraction speaking English	-0.0078	-	-0.0074	-0.0064	-
Fraction speaking foreign language	-	-	-0.0002	0.0013	-
Exchange rate distortions	-	-	1.1×10^{-6}	-1.1×10^{-5}	-
Equipment investment	0.1511	0.2181	-	0.0917	-
Non-equipment investment	0.0295	-	0.0575	0.0551	-
Std dev of black market premium	-	-	-4.5×10^{-6}	-2.0×10^{-6}	-
Outward orientation	-0.0035	-	-0.0023	-0.0021	-
Black market premium	-0.0055	-	-0.0057	-0.0047	-
Total area of the country	-	-	-1.2×10^{-7}	-7.3×10^{-7}	-3.4×10^{-7}
Latin American dummy	-0.0127	-	-0.0083	-0.0046	-0.0134
Sub-Saharan dummy	-0.0218	-	-0.0204	-0.0184	-0.0259
Higher education enrollment	-0.1213	-	-	-	-
Public education share	-	-	-	-	-
Revolutions and coups	-	-	0.0017	-0.0009	-
War dummy	-	-	-0.0045	-0.0043	-
Political rights	-	-	-0.0020	-0.0017	-
Civil liberties	-0.0028	-	0.0006	-5.8×10^{-5}	-0.0024
Absolute latitude	-	-	-0.0001	1.2×10^{-5}	-0.0002
Average age of the population	-	-	-2.7×10^{-5}	-4.2×10^{-5}	-
British colony dummy	0.0079	-	0.0012	0.0012	-
Fraction of buddhist	-	-	0.0074	0.0072	-
Fraction of catholic	-	-	-0.0026	-0.0031	-
Fraction of confucius	0.0759	0.0577	0.0537	0.0592	0.0539
Ethnolinguistic fractionalization	0.0165	-	-	0.0089	-
French colony dummy	0.0110	-	-	-	-
Fraction of hindu	-0.1108	-	-0.0600	-0.0694	-0.0402
Fraction of jewish	-	-	-	-	-
Fraction of muslim	0.0078	0.0118	0.0039	0.0024	-
Primary exports	-	-	-0.0047	-0.0003	-
Fraction of protestant	-	-0.0136	-0.0094	-0.0097	-
Rule of law	0.0131	-	0.0099	0.0073	0.0199
Spanish colony dummy	0.0140	-	-	-	-
Growth rate of population	-	-	-	-	-
Ratio workers to population	-	-	-0.0053	-0.0062	-
Labor force	3.8×10^{-7}	-	1.7×10^{-7}	2.1×10^{-7}	5.8×10^{-8}
R^2	95%	81%	92%	93%	77%
R^2(adjusted)	92%	79%	86%	87%	72%

Chapter 6

Conclusion

Understanding factors that form comparative advantages and identifying their stimulation measures are challenging tasks. In the above papers I demonstrate that such analyses should be based both on a sound theoretical framework and advanced econometric methods. I show that failures in one of those areas may either lead to controversial results (which are inconsistent with empirical findings) or cause biased estimates (due to, e.g., the 'Rashomon Effect'), even though the obtained regressions seem impeccable from a technical point of view. In this regard, I confirm the knowledge that economic theory and econometrics should go hand in hand opening new vistas in the areas of macroeconomics.

Analyzing factors forming comparative advantages between countries based on their competitive advantages, I consider prices under trade distortions as approximations to autarky prices that will converge to international prices after economic liberalization. This effectively restricts our analysis to economies being in the period of transition, but, as I show above, provides an efficient tool forecasting comparative advantages' dynamics, which is then fortified on the example of the CEE countries. Applying this tool to the Russian data, I identify a series of medium and high technological industries and sub-industries that exhibit potential comparative advantages including pharmaceutical products, electronic equipment, machinery building and railway transport.

Developing a comprehensive theory that connects macro– and micro–level of analysis, I show that the capital–biased technical change observed empirically in a number of developing and transition economies over the last decades was triggered by capital–market integration and resulting massive inflow of FDI. Considering new technologies, that increase factor productivity, to be endogenously driven and defined by companies, which provide those technologies in the sector with higher expected profits, I identify some general ways to stimulate comparative advantages. In particular, those are attraction of (scarce) capital and mitigation of a potential inefficiency on the market of new technologies.[1]

In an attempt to explore factors that stimulate companies to innovate in more detail, I suggest and estimate a generalized model based on regional data from Russia. Dealing with the problem of optimal model selection, unobserved heterogeneity and possible endogeneity of regressors, I determine a series of factors having a significant impact on the innovative output of companies discussing reasons, which explain those findings, and potential implications for industrial policy.

Taking into account the current modernization program of the Russian economy with its long-term strategy oriented towards innovations in industries with high value–added goods (e.g., 'Federal Targeted Programme of Nanotechnologies'), the findings stated can be considered as illustrative for policy makers promoting further acceptance and usage of econometric tools in policy–relevant analysis. Furthermore, since no specific assumptions (being true exclusively for the Russian economy) are taken, the methods developed in my thesis can be well applied later to other BRIC members and economies in transition.

Apart from an in-depth theoretical analysis and a well structured empirical examination, availability and quality of data is found to be the major challenge for this work. While in certain cases no good solution is identified (e.g., availability of disaggregated microdata in the Russian regional panel dataset for testing particular hypotheses) leaving this for future research, limited data availability (e.g., short time series dimension) can be considered as a general problem for economies in transition including the CEE countries and Russia.

This problem becomes particularly demanding if complex dynamic models with instrumental variables or large multivariate lag structures need to be estimated. To relax this requirement, I apply heuristic optimization methods in my thesis as a powerful instrument of robust model selection that allows one to reduce variance in results and avoid inconsistent estimates. Being highly interested in further application of these methods, I conduct two more studies. Thus, I find evidence on their effectiveness for vector autoregressive models with short time series dimensions. Furthermore, comparing heuristic strategies with the Lasso–type subset selection methods, I identify conditions required for each of the two strategies to demonstrate a good performance.

With this work I believe I contribute to the existing research practice bringing different instruments of economic analysis closer together and giving more comprehensive understanding of real economic processes.

[1] Unfortunately, the corresponding chapter is not included in the book, but can be found in MAGKS Joint Discussion Paper Series in Economics 18-2011 at http://www.uni-marburg.de/fb02/makro/forschung/magkspapers/18-2011_savin.pdf.

Bibliography

Acemoglu, D. (2002). Directed technical change. *Review of Economic Studies* **69**, 781–809.

Acemoglu, D., P. Aghion and F. Zilibotti (2003). Vertical integration and distance to frontier. *Journal of the European Economic Association* **1**(2-3), 630–638.

Acosta-Gonzalez, E. and F. Fernandez-Rodriguez (2007). Model selection via genetic algorithms illustrated with cross-country growth data. *Empirical Economics* **33**(2), 313–337.

Adam, A., T.S. Kosma and J. McHugh (2003). Trade liberalization strategies: What could southeastern europe learn from CEFTA and BFTA. Technical Report 239. IMF Working Paper.

Aghion, P. and P. Howitt (1992). A model of growth through creative destruction. *Econometrica* **60**(2), 323–351.

Agrawal, M. (1999). *Global Competitiveness in the Pharmaceutical Industry: the Effect of National Regulatory, Economic and Market Factors*. Haworth Press.

Alberto, I., A. Beamonte, P. Gargallo, P.M. Mateo and M. Salvador (2010). Variable selection in STAR models with neighbourhood effects using genetic algorithms. *Journal of Forecasting* **29**(8), 728–750.

Arellano, M. and S. Bond (1991). Some tests of specification for panel data: Monte Carlo evidence and an application to employment equations. *Review of Economic Studies* **58**(2), 277–297.

Armagan, A. and R. L. Zaretzki (2010). Model selection via adaptive shrinkage with t priors. *Computational Statistics* **25**(3), 441–461.

Aukutsionek, S.P. and A.V. Belianin (2001). Quality of forecasts and business performance: A survey study of Russian managers. *Journal of Economic Psychology* **22**(5), 661–692.

Bacchiocchi, E. and F. Montobbio (2009). Knowledge diffusion from university and public research. a comparison between US, Japan and Europe using patent citations. *Journal of Technology Transfer* **34**(2), 169–181.

Bach, F. (2008). Bolasso: model consistent Lasso estimation through the bootstrap. In: *Proceedings of the Twenty-fifth International Conference on Machine Learning.*

Bach, F. and G. Obozinski (2010). *Sparse methods for machine learning: Theory and algorithms. Tutorial of ECML.* http://www.di.ens.fr/~fbach/ecml2010tutorial/. Accessed 12 September 2011.

Backe, P. (1999). Exchange rate regimes in Central and Eastern Europe: A brief review of recent changes, current issues and future challenges. Technical report.

Balance, R.H., H. Forstner and T. Murray (1987). Consistency tests of alternative measures of comparative advantage, the review of economics and statistics. *The Review of Economics and Statistics* **69**(1), 157–161.

Balassa, B. (1965). Trade liberalization and "revealed" comparative advantage. *The Manchester School of Economic and Social Studies* (33), 92–123.

Balk, B.M. (1983). Does there exist a relation between inflation and relative price change variability?. *Economic letters* **13**(2–3), 173–180.

Bardhan, A.D. and C. Kroll (2006). Competitiveness and an emerging sector: The russian software industry and its global linkages. *Industry and Innovation* **13**(1), 69–95.

Belousov, A. R. (2006). The scenarios of russian economic development for the next 15 years (in Russian). *"Problemj Prognosirovanija"* **13**(1), 3–53.

Blundell, R. and S. Bond (1998). Initial conditions and moment restrictions in dynamic panel data models. *Journal of Econometrics* **87**(1), 115–143.

Blundell, R., R. Griffith and J. Van Reenen (1999). Market share, market value and innovation in a panel of British manufacturing firms. *Review of Economic Studies* **66**, 529–554.

Box, G.E.P. and D.A. Pierce (1970). Distribution of residual autocorrelation in autoregressive moving average time series models. *Journal of the American Statistical Association* **65**(332), 1509–1526.

Breiman, L. (2001). Statistical modelling: the two cultures. *Statistical Science* **16**(3), 199–231.

Brenton, P., N. Tourdyeva and J. Whalley (1997). The potential trade effects of an FTA between the EU and Russia. *Review of World Economics* **133**(2), 205–225.

Bruno, M. (1965). The optimal selection of export-promoting and import-substituting projects. In: *Planning the External Sector: Techniques, Problems and Policies* (United Nations, Ed.). New York.

Bucci, A. and C. P. Parello (2009). Horizontal innovation-based growth and product market competition. *Economic Modelling* **26**(1-2), 213–221.

Cainelli, G., R. Evangelista and M. Savona (2006). Innovation and economic performance in services: a firm-level analysis. *Cambridge Journal of Economics* **30**(3), 435–458.

Candès, E. J. and T. Tao (2007). The Dantzig selector: statistical estimation when p is much larger than n. *Annals of Statistics* **35**(6), 2313–2351.

Candès, E. J., B. Wakin and S. Boyd (2008). Enhancing sparsity by reweighted l1 minimization. *Journal of Fourier Analysis and Applications* **14**, 877–905.

Caner, M. (2009). Lasso-type GMM estimator. *Econometric Theory* **25**, 270–290.

Carstensen, K., K. Wohlrabe and C. Ziegler (2011). Predictive ability of business cycle indicators under test. A case study for the Euro Area industrial production. *Journal of Economics and Statistics* **231**(1), 82–106.

Chamberlin, G. (2007). Forecasting GDP using external data sources. *Economic & Labour Market Review* **1**(8), 18–23.

Claveria, O., E. Pons and R. Ramos (2007). Business and consumer expectations and macroeconomic forecasts. *International Journal of Forecasting* **23**(1), 47–69.

Crescenzi, R., A. Rodriguez-Posa and M. Storper (2007). The territorial dynamics of innovation: a Europe-United States comparative analysis. *Journal of Economic Geography* **7**(6), 673–709.

Dezhina, I. G., N. V. Shelubinskaya, L. K. Pipiya and N. I. Ivanova (2008). Innovation policy analysis and appraisal of its results (in Russian). *Innovations* **5**(115), 56–72.

Dickey, D.A. and W.A. Fuller (1979). Distribution of the estimators for autoregressive time series with a unit root. *Journal of the American Statistical Association* **74**(366), 427–431.

Diebold, F. X. and R. S. Mariano (1995). Comparing predictive accuracy. *Journal of Business & Economic Statistics* **13**(3), 253–263.

Diewert, W.E (1993). The economic theory of index numbers: A survey. In: *Essays in Index Number Theory* (Diewert W.E and A.O. Nakamura, Eds.). Vol. 1. pp. 163–208. Amsterdam: North-Holland.

Dixit, A. K. and V. Norman (1989). *Theory of International Trade: a Dual, General Equilibrium Approach*. Cambridge University Press.

Domsch, M.E. and T. Lidokhover (2007). The broader historical, social and economic context of the current situation in Russia. In: *Human Resource Management in Russia* (M.E. Domsch and T. Lidokhover, Eds.). pp. 1–22. Aldershot. Hampshire, UK.

Dornbusch, R., S. Fischer and P.A. Samuelson (1977). Comparative advantage, trade and payments in a ricardian model with a continuum of goods. *American Economic Review* **67**(5), 823–839.

Dorsey, R.E. and W.J. Meyer (1995). Genetic algorithms for estimation problems with multiple optima, nondifferentiability, and other regular features. *Journal of Business and Economic Statistics* **13**(1), 53–66.

Drechsel, K. and R. Scheufele (2010). Should we trust in leading indicators? Evidence from the recent recession. Technical Report 10-10. IWH Discussion Papers.

Durand, M. and C. Giorno (1987). Indicators of international competitiveness: Conceptual aspects and evaluation. Technical Report 9. OECD Economic Studies.

Efron, B., T. Hastie, I. Johnstone and R. Tibshirani (2004). Least angle regression. *Annals of Statistics* **32**, 407–489.

Fan, J. and J. Lv (2008). Sure independence screening for ultrahigh dimensional feature space. *Journal of the Royal Statistical Society B* **70**(5), 849–911.

Fernandez, C., E. Ley and M. Steel (2001). Model uncertainty in cross-country growth regressions. *Journal of Applied Econometrics* **16**, 563–576.

Figueiredo, M. A. T., R. D. Nowak and S. J. Wright (2007). Gradient projection for sparse reconstruction: application to compressed sensing and other inverse problems. *IEEE Journal of Selected Topics in Signal Processing: Special Issue on Convex Optimization Methods for Signal Processing* **1**(4), 586–598.

Flaig, G. (2003). Seasonal and cyclical properties of ifo business test variables. *Journal of Economics and Statistics* **223**(5), 556–570.

Fogel, D. B. (2006). *Evolutionary Computation: Toward a New Philosophy of Machine Intelligence*. Wiley-IEEE Press. Hoboken, NJ.

Forrest, J. E. (1991). Models of the process of technological innovation. *Technology Analysis & Strategic Management* **3**(4), 439–453.

Foster, S. D., A. P. Verbyla and W. S. Pitchford (2008). A random model approach for the lasso. *Computational Statistics* **23**, 217–233.

Frank, I. E. and J. H. Friedman (1993). A statistical view of some chemometrics regression tools. *Technometrics* **35**(2), 109–135.

Friedman, J., T. Hastie, H. Höfling and R. Tibshirani (2007). Pathwise coordinate optimization. *The Annals of Applied Statistics* **1**(2), 302–332.

Funk, M. (2006). Business cycles and research investment. *Applied Economics* **38**, 1775–1782.

Gasso, G., A. Rakotomamonjy and S. Canu (2009). Recovering sparse signals with a certain family of non-convex penalties and DC programming. *IEEE Transactions on Signal Processing* **57**(12), 4686–4698.

Gatu, C., E. J. Kontoghiorghes, M. Gilli and P. Winker (2008). An efficient branch-and-bound strategy for subset vector autoregressive model selection. *Journal of Economic Dynamics & Control* **32**, 1949–1963.

Gerschenkron, A. (1962). *Economic Backwardness in Historical Perspective: A Book of Essays*. Cambridge, MA: Belknap Press of Harvard University Press.

Giacomini, R. and H. White (2006). Tests of conditional predictive ability. *Econometrica* **74**(6), 1545–1578.

Gilli, M. and P. Winker (2009). Heuristic optimization methods in econometrics. In: *Handbook of Computational Econometrics* (D. A. Belsley and E. Kontoghiorghes, Eds.). pp. 81–119. Wiley. Chichester.

Goggin, J. (2008). An analysis of the potential of the European Commission business and consumer surveys for macroeconomic forecasting. *ESRI Quarterly Economic Commentary* **4**, 46–67.

Goldrian, G. (2007). A leading indicator composed of survey data: appropriate construction and prognostic significance. In: *Handbook of Survey-Based Business Cycle Analysis* (G. Goldrian, Ed.). pp. 151–158. Edward Elgar. Cheltenham.

Grossman, G. and E. Helpman (1991). *Innovation and Growth in the Global Economy*. Cambridge: MIT Press.

Hastie, T., R. Tibshirani and J. Friedman (2009). *The Elements of Statistical Learning: Data Mining, Inference, and Prediction*. Springer Verlag. New York.

Hausman, J. A. and W. E. Taylor (1981). Panel data and unobservable individual effects. *Econometrica* **49**(6), 1377–1398.

Hendry, D. F. and H. M. Krolzig (2005). The properties of automatic "GETS" modelling. *The Economic Journal* **115**(502), C32–C61.

Hsu, N.-J., H.-L. Hung and Y.-M. Chang (2007). Subset selection for vector autoregressive processes using lasso. *Computational Statistics & Data Analysis* **52**(7), 3645–3657.

Hüfner, F. and M. Schröder (2002). Forecasting German industrial production: An econometric comparison of ifo- and ZEW-business expectations. *Journal of Economics and Statistics* **222**(3), 316–336.

Jefferson, G., A. G. Z. Hu, X. Guan and X. Yu (2003). Ownership, performance, and innovation in China's large- and medium-size industrial enterprise sector. *China Economic Review* **14**(1), 89–113.

Johnson, B. A. (2009). On lasso for censored data. *Electronic Journal of Statistics* **3**, 485–506.

Kapetanios, G. (2007). Variable selection in regression models using nonstandard optimization of information criteria. *Computational Statistics & Data Analysis* **52**(1), 4–15.

Kapetanios, G., V. Labhard and S. Price (2008). Forecasting using Bayesian and information-theoretic model averaging: An application to U.K. inflation. *Journal of Business & Economic Statistics* **26**(1), 33–41.

Kaucic, M. (2009). Predicting EU energy industry excess returns on EU market index via a constrained genetic algorithm. *Computational Economics* **34**, 173–193.

Kholodilin, K. A. and B. Siliverstovs (2006). On the forecasting properties of the alternative leading indicators for the German GDP: Recent evidence. *Journal of Economics and Statistics* **226**(3), 234–259.

Klotz, S. (1997). Econometric models with spatial autocorrelation - an introductory survey. *Journal of Economics and Statistics* **218**(1+2), 168–196.

Knight, K. and W. Fu (2000). Asymptotics for lasso-type estimators. *Annals of Statistics* **28**(5), 1356–1378.

Krugman, P. R. and G. N. Hatsopoulos (1987). The problem of U.S. competitiveness in manufacturing. *New-England Economic Review* pp. 18–29.

Kwiatkowski, D., P.C.B. Phillips, P. Schmidt and Y. Shin (1992). Testing the null hypothesis of stationarity against the alternative of a unit root. *Journal of Econometrics* **54**(1-3), 159–178.

Lafay, G. (1992). The measurement of revealed comparative advantage. In: *International Trade Modelling* (M. G. Dagenais and P. A. Muet, Eds.). pp. 209–234. London: Chapman & Hall.

Landesmann, M. A. and R. Stehrer (2001). Convergence patterns and switchovers in comparative advantage. *Structural Change and Economic Dynamics* **12**, 399–423.

Langenmantel, E. (1999). Das ifo-Geschäftsklima als Indikator für die Prognose des Bruttoinlandproduktes. *IFO Schnelldienst* **16/17**, 16–21.

Lanne, M., H. Lütkepohl and P. Saikkonen (2002). Comparison of unit root tests for time series with level shifts. *Journal of Time Series Analysis* **23**(3), 667–685.

Lütkepohl, H. (2007). *New Introduction to Multiple Time Series Analysis. Corr. 2nd printing*. Springer. Berlin, Heidelberg, New York.

Lv, B. J. and Y. Fan (2009). A unified approach to model selection and sparse recovery using regularized least squares. *The Annals of Statistics* **37**(6A), 3498–3528.

MacGarvie, M. (2001). The determinants of international knowledge diffusion as measured by patent citations. *Economics Letters* **87**, 121–126.

Marconi, D. and V. Rolli (2008). Comparative advantage patterns and domestic determinants in emerging countries. Technical Report 81. World Institute for Development Economics Research, Research Paper.

Maringer, D. and P. Winker (2009). The convergence of estimators based on heuristics: Theory and application to a GARCH model. *Computational Statistics* **24**, 533–550.

Meinshausen, N. and B. Yu (2008). Lasso-type recovery of sparse representations for high-dimensional data. *Annals of Statistics* **37**(1), 246–270.

Merivate, E. J. and J. C. Pernias (2006). Innovation complementarity and scale of production. *Journal of Industrial Economics* **54**(1), 1–29.

Meyer, M. and P. Winker (2005). Using HP filtered data for econometric analysis: some evidence from Monte Carlo simulations. *Allgemeines Statistisches Archiv* **89**(3), 303–320.

Mincer, J. A. and V. Zarnowitz (1969). The evaluation of economic forecasts. In: *Economic Forecasts and Expectations: Analysis of Forecasting Behavior and Performance* (J. A. Mincer, Ed.). pp. 1–46. NBER. New York.

Mittnik, S. and P. Zadrozny (2005). Forecasting quarterly German GDP at monthly intervals using monthly ifo business conditions data. In: *Ifo Survey Data in Business Cycle Analysis and Monetary Policy Analysis* (J.-E. Sturm and T. Wollmershï¿½user, Eds.). pp. 19–49. Physica-Verlag. Heidelberg.

Montobbio, F. (2003). Sectoral patterns of technological activity and export market share dynamics. *Cambridge Journal of Economics* **27**, 523–545.

Nakamura, E. and J. Steinsson (2008). Five facts about prices: A reevaluation of menu cost models. *The Quarterly Journal of Economics* **123**(4), 1415–1464.

Nardo, M. (2003). The quantification of qualitative business surveys: a critical assessment. *Journal of Economic Surveys* **47**(2), 211–223.

Nilsson, R. and O. Brunet (2006). Composite leading indicators for major OECD non-member economies: Brazil, China, India, Indonesia, Russian Federation, South Africa. Technical Report 2006/1. OECD Statistics Working Papers.

OECD (1986). Technological performance and industrial competitiveness. Technical Report 2. Science and Technology Indicators.

OECD (2003). *Business Tendency Surveys: A Handbook*. OECD Organisation for Economic Co–operation and Development. http://www.oecd.org/dataoecd/29/61/31837055.pdf.

Okui, R. (2009). The optimal choice of moments in dynamic panel data models. *Journal of Econometrics* **151**(1), 1–16.

Osborne, M., B. Presnell and B. Turlach (2000). On the lasso and its dual. *Journal of Computational and Graphical Statistics* **9**(2), 319–337.

Ozyldrim, A., B. Schaitkin and V. Zarnowitz (2010). Business cycles in the Euro area defined with coincident economic indicators and predicted with leading economic indicators. *Journal of Forecasting* **29**(1-2), 6–28.

Penrose, R. (1956). On best approximate solutions of linear matrix equations. *Proceedings of the Cambridge Philosophical Society* **52**(1), 17–19.

Perez-Amaral, T., G. M. Gallo and H. White (2003). A flexible tool for model building: The relevant transformation of the inputs network approach (RETINA). *Oxford Bulletin of Economics and Statistics* **65**(1), 821–838.

Polterovich, V. M. (2009). The problem of creating a National Innovation System (in Russian). *Economics and Mathematical Methods* **45**(2), 3–18.

Porter, M. (1990). *The Competitive Advantage of Nations*. Harvard Business Review.

Porter, M. E. (2003). The economic performance of regions. *Regional Studies* **37**(6-7), 549–578.

Rautava, J. (2005). *BOFIT Russia Review Yearbook 2005*. Bank of Finland. Helsinki.

Reeves, C. and J.F. Rowe (2003). *Genetic Algorithms - Principles and Perspectives*. Kluwer. Boston, MA.

Rogov, V. V. (2004). Russian export potential: Current state, lines and conditions of development (in Russian). *"Vneshneekonomicheskij Bulleten"* (5), 61–83.

Romer, P. M. (1990). Endogenous technological change. *Journal of Political Economy* **98**(5), 71–102.

Rosenberg, N. R., R. Landau and D. C. Mowery (1992). *Technology and the Wealth of Nations*. Stanford University Press. Stanford, CA.

Sala-i-Martin, X. (1997). I just ran two million regressions. *American Economic Review* **87**(2), 178–183.

Savin, I. (forthcoming). A comparative study of the lasso-type and heuristic model selection methods. *Journal of Economics and Statistics*.

Savin, I. and P. Winker (2011). Heuristic model selection for leading indicators in Russia and Germany. Technical Report 01-2011. MAGKS Joint Discussion Paper Series in Economics.

Savin, I. and P. Winker (2012). Heuristic optimization methods for dynamic panel data model selection. Application on the Russian innovative performance. *Computational Economics* **39**(4), 337–363.

Savin, I. and P. Winker (forthcoming). Lasso-type and heuristic strategies in model selection and forecasting. In: *Towards Advanced Data Analysis by Combining Soft Computing and Statistics* (C. Borgelt, M.A. Gil, J. Sousa and M. Verleysen, Eds.). Springer. Berlin.

Schneider, U. and M. Wagner (2009). Catching growth determinants with the adaptive lasso. Technical Report 55. wiiw Working Paper.

Schumpeter, J. A. (1943). *Capitalism, Socialism and Democracy*. Allen Unwin. London.

Shibata, R. (1976). Selection of the order of an autoregressive model by Akaike's information criterion. *Biometrika* **63**(1), 117–126.

Shinohara, K. and H. Okuda (2010). Dynamic innovation diffusion modelling. *Computational Economics* **35**, 51–62.

Siggel, E. (2006). International competitiveness and comparative advantage: A survey and a proposal for measurement. *Journal of Industry, Competition and Trade* **6**(2), 137–159.

Sin, C.-Y. and H. White (1996). Information criteria for selecting possibly misspecified parametric models. *Journal of Econometrics* **71**(1-2), 207–225.

Sölvell, Ö. (2008). *Clusters - Balancing Evolutionary and Constructive Forces*. Ivory Tower Publishers. Stockholm.

Stehrer, R. and J. Woerz (2003). Technological convergence and trade patterns. *Review of World Economics* **139**, 191–219.

Strigel, W. (1965). Die Konjunkturumfragen des ifo-Instituts für Wirtschaftsforschung. *Allgemeines Statistisches Archiv* **49**, 129–163.

Strigel, W. (1997). Seit 25 Jahren: Das ifo Geschäftsklima als Konjunkturindikator. *ifo Schnelldienst* **27**, 3–4.

Swansan, N. R. (1998). Money and output viewed through a rolling window. *Journal of Monetary Economics* **41**(3), 455–474.

Tarr, D.G. (2007). The structure of import tariffs in the Russian Federation: 2001-05. Technical Report 4265. World Bank Policy Research.

Theil, H. (1966). *Applied Economic Forecasting*. Rand McNally. Chicago.

Tibshirani, R. (1996). Regression shrinkage and selection via the lasso. *Journal of the Royal Statistical Society B* **58**(1), 267–288.

Tiusanen, T. (2007). Development of Rouble Exchange Rate in Russia. Technical Report 45. NORDI series. Lappeenranta.

Tsukhlo, S. V. (2005). *The IET surveys, organisational problems and analysis opportunities (in Russian)*. Institute for the Economy in Transition. http://www.iet.ru/files/persona/tsukhlo/council_24.03.05.pdf.

Turlach, B. A., W. N. Venables and S. J. Wright (2005). Simultaneous variable selection. *Technometrics* **47**(3), 349–363.

Tytell, I. and K. Yudaeva (2006). The role of FDI in Eastern Europe and New Independent States. Technical report. Centro Studi Luca d'Agliano Development Studies Working Paper No. 217. Milan.

Vinogradov, B. A. (2008). Certain aspects of staffing for defensive-industrial complex (in Russian). *Innovations* **7**(117), 23–29.

West, K. D. (1996). Asymptotic inference about predictive ability. *Econometrica* **64**(5), 1067–1084.

Winker, P. (1995). Identification of multivariate AR-models by threshold accepting. *Computational Statistics & Data Analysis* **20**, 295–307.

Winker, P. (1999). Causes and effects of financing constraints at the firm level: Some microeconometric evidence. *Small Business Economics* **12**, 169–181.

Winker, P. (2000). Optimized multivariate lag structure selection. *Computational Economics* **16**, 87–103.

Winker, P. (2001). *Optimization Heuristics in Econometrics: Applications of Threshold Accepting*. Wiley. Chichester.

Woerz, J. (2005). Dynamics of trade specialization in developed and less developed countries. *Emerging Markets, Finance and Trade* **41**, 92–111.

WTO (1997). Cyprus: June 1997. Technical report. Trade Policy Reviews: First Press Release, Secretariat and Government Summaries.

Yuan, M. and Y. Lin (2007). On the non-negative garrotte estimator. *Journal of the Royal Statistical Society B* **69**(2), 143–161.

Zaghini, A. (2005). Evolution of trade patterns in the new EU member states. *Economics of Transition* **13**, 629–658.

Zhang, T. (2010). Analysis of multi-stage convex relaxation for sparse regularization. *Journal of Machine Learning Research* **11**, 1081–1107.

Zhao, P. and B. Yu (2006). On model selection consistency of lasso. *Journal of Machine Learning Research* **7**, 2541–2563.

Zou, H. (2006). The adaptive lasso and its oracle properties. *Journal of the American Statistical Association* **101**, 1418–1429.

Zou, H. and H.H. Zhang (2009). On the adaptive elastic-net with a diverging number of parameters. *Annals of Statistics* **37**(4), 1733–1751.

Zou, H. and T. Hastie (2005). Regularization and variable selection via the elastic net. *Journal of the Royal Statistical Society B* **67**(2), 301–320.

Schriften zur empirischen Wirtschaftsforschung

Herausgegeben von Peter M. Schulze und Peter Winker

Band 1 Christoph Balz: Multivariate Überprüfung von Hysteresiseffekten. Eine empirische Analyse ausgewählter Arbeitsmärkte. 1999.

Band 2 Daniel Porath: Fiskalische Beurteilung der Staatsverschuldung mit ökonometrischen Methoden. Eine empirische Studie für die Bundesrepublik Deutschland. 1999.

Band 3 Martina Johannsen: Theorie und Empirie von Arbeitsmärkten. Eine ökonometrische Analyse für Rheinland-Pfalz. 2000.

Band 4 Peter M. Schulze: Regionales Wachstum und Strukturwandel. Quantitative Analyse mit Regionaldaten für die Bundesrepublik Deutschland. Unter Mitarbeit von Christoph Balz. 2001.

Band 5 Nora Lauterbach: Tertiarisierung und Informatisierung in Europa. Eine empirische Analyse des Strukturwandels in Deutschland, Frankreich, Italien und Großbritannien. 2004.

Band 6 Robert Skarupke: Renditen von Bildungsinvestitionen. Paneldaten-Schätzungen für die Bundesrepublik Deutschland. 2005.

Band 7 Manfred Scharein: Zur Theorie skalenparametergesplitteter Verteilungen und ihrer Anwendung auf den deutschen Aktienmarkt. 2005.

Band 8 Ke Ma: Quantitative Renditeanalysen am deutschen Aktienmarkt mit Multifaktoren-Modellen. 2005.

Band 9 Jens Ulrich Hanisch: Rounding of Income Data. An Empirical Analysis of the Quality of Income Data with Respect to Rounded Values and Income Brackets with Data from the European Community Household Panel. 2007.

Band 10 Yvonne Lange: Fertilität und Erwerbsbeteiligung von Frauen in Deutschland. Eine empirische Analyse. 2007.

Band 11 Jörg Schmidt: Relative Deprivation, Arbeitszufriedenheit und Betriebswechsel. Eine Analyse auf Basis von Linked Employer-Employee Daten. 2008.

Band 12 Tanja Kasten: Monetäre und nicht-monetäre Effekte von Erwerbsunterbrechungen. Eine mikroökonometrische Analyse auf Basis des SOEP. 2008.

Band 13 Alexander Prinz: Quantitative Analysen zum deutschen und internationalen Luftfrachtmarkt. 2008.

Band 14 Anke Koch: Regionale Verteilung der Beschäftigung in Deutschland. Panel- und Zähldatenmodelle. 2009.

Band 15 Martin Mandler: Geldpolitische Reaktionsfunktionen und makroökonomische Unsicherheit. 2010.

Band 16 Verena Dexheimer: Einreisetourismus in Deutschland. Paneldatenanalysen und SARIMA-Prognosen. 2010.

Band 17 Julia König: Gesundheitsausgaben in Deutschland. Eine Kointegrationsanalyse. 2010.

Band 18 Akwum Onwunta: Contributions to Credit Portfolio Modeling and Optimization. 2011.

Band 19 Vahidin Jeleskovic: Agentenbasierte Modelle für empirische Wechselkurse. Ökonometrische Schätzung und Evaluierung. 2011.

Band 20 Jörg Böttcher: Möglichkeiten einer Projektfinanzierung bei CSP-Vorhaben. 2012.

Band 21 Ivan Savin: Russia´s Comparative Advantages in Foreign Trade. Forecasting Potential Dynamics and Identifying Stimulation Measures. 2013.

www.ingramcontent.com/pod-product-compliance
Ingram Content Group UK Ltd.
Pitfield, Milton Keynes, MK11 3LW, UK
UKHW021817051026
15470UKWH00026B/1394

9 783631 639931